JOURNAL OF ICT STANDARDIZATION

Volume 1, No. 3 (March 2014)

JOURNAL OF ICT STANDARDIZATION

Chairperson: Ramjee Prasad, CTIF, Aalborg University, Denmark
Editor-in-Chief: Anand R. Prasad, NEC, Japan
Advisors: Bilel Jamoussi, ITU, Switzerland
Jesper Jerlang, Dansk Standard, Denmark

Editorial Board
Kiritkumar Lathia, Independent ICT Consultant, UK
Hermann Brandt, ETSI, France
Kohei Satoh, ARIB, Japan
Sunghyun Choi, Seoul National University, South Korea
Ashutosh Dutta, AT&T, USA
Alf Zugenmaier, University of Applied Sciences Munich, Germany
Julien Laganier, Juniper Networks, USA
John Buford, Avaya, USA
Monique Morrow, Cisco, Switzerland
Vijay K. Gurbani, Alcatel Lucent, USA
Henk J. de Vries, Rotterdam School of Management, Erasmus University, The Netherlands
Yoichi Maeda, TTC Japan
Debabrata Das, IIIT-Bangalore, India
Signe Annette Bøgh, Dansk Standard, Denmark
Rajarathnam Chandramouli, Stevens Institute of Technology, USA

Objectives:

- Bring papers on de-jure as well as de-facto standards to the readers
- Cover pre-development, including technologies with potential of becoming a standard, as well as developed / deployed standards
- Publish on-going work with potential of becoming a standard technology
- Publish papers giving explanation of standardization process
- Publish tutorial type papers giving new comers a understanding of standardization

Aims & Scope

- Aim:
  - The aim of this journal is to publish standardized as well as related work making "standards" accessible to a wide public – from practitioners to new comers.
  - The journal aims at publishing in-depth as well as overview work including papers discussing standardization process and those helping new comers to understand how standards work.
- Scope:
  - Bring up-to-date information regarding standardization in the field of Information and Communication Technology (ICT) covering all protocol layers and technologies in the field

JOURNAL OF ICT STANDARDIZATION

Volume 1, No. 3 (March 2014)

KOHEI SATOH / The R & D and Standardization Activities of ARIB	271–286
TOSHIAKI KUROKAWA / New Approaches for Task Classification about Standardization Skills	287–300
DAVID LAKE, RODOLFO MILITO, MONIQUE MORROW AND RAJESH VARGHEESE / Internet of Things: Architectural Framework for eHealth Security	301–328
RAHAMATULLAH KHONDOKER, ABBAS SIDDIQUI, PAUL MÜLLER AND KPATCHA BAYAROU / Realization of Service-Orientation Paradigm in Network Architectures	329–346
ARTURO SERRANO-SANTOYO, VERONICA ROJAS-MENDIZABA, ROBERTO CONTE-GALVAN, AMANDA GOMEZ-GONZALEZ AND ANGELICA BAPTISTA SILVA / Towards a Framework for Health IT Standardization in Mexico	347–362
Author Index	363–364
Keywords Index	365–366

Published, sold and distributed by:
River Publishers
PO box 1657
Algade 42
9000 Aalborg
Denmark
Tel.: +4536953197

www.riverpublishers.com

Journal of ICT Standardization is published three times a year. Publication programme, 2013–2014: Volume 1 (3 issues)

ISSN: 2245-800X
ISSN: 978-87-93102-79-8

All rights reserved © 2014 River Publishers

No part of this publication may be reproduced, stored in a retrieval system, or transmitted in any form or by any means, mechanical, photocopying, recording or otherwise, without prior written permission of the publishers.

The R & D and Standardization Activities of ARIB

Kohei Satoh

Managing Director, Association of Radio Industries and Businesses (ARIB)

Received: 29 March, 2013; Accepted: 9 October, 2013

Abstract

Radio systems in the field of telecommunication and broadcasting, as seen in mobile phones, wireless LAN, ITS, digital broadcasting, have been increasingly sophisticated and diversified with the rapid advances in digital technologies. They are essential for our economic activities and social life. Research and development (R & D) and standardization for new radio systems have become more important, so that the role of the Association of Radio Industries and Businesses (ARIB), as a standards development organization (SDO) for radio systems, has been increasing. The paper describes the current overview and future challenges of ARIB.

1 ARIB

ARIB was established in response to several trends, such as the growing internationalization of telecommunications, the convergence of telecommunications and broadcasting, and the need for the promotion of radio-related industries. ARIB's goal is to rapidly advance the use of radio technology for the benefit of society. This is done by integrating knowledge and experience in the various fields of radio use, such as broadcasting and telecommunications, research & development (R & D) in radio technology, and by serving as a standards development organization (SDO) for radio technology.

ARIB was chartered by the Minister of Posts and Telecommunication as a public service corporation on May 15, 1995. Its activities include those previously performed by the Research and Development Center for Radio

System (RCR) and the Broadcasting Technology Association (BTA). ARIB renewed its organization based on the General Incorporated Association and General Incorporated Foundations Act in April 2011.

The main activities of ARIB are as follows:

(1) Study and R & D on the utilization of radio waves in the field of telecommunications and broadcasting,
(2) Consultation, enlightenment for dissemination, and collection and provision of material or information on the utilization of radio waves in the field of telecommunications and broadcasting,
(3) Development of standards relating to radio systems in the field of telecommunications and broadcasting, and
(4) Liaison, coordination and cooperation with foreign organizations relating to regarding the utilization of radio waves in the field of telecommunications and broadcasting.

Figure 1 shows the organization of ARIB.

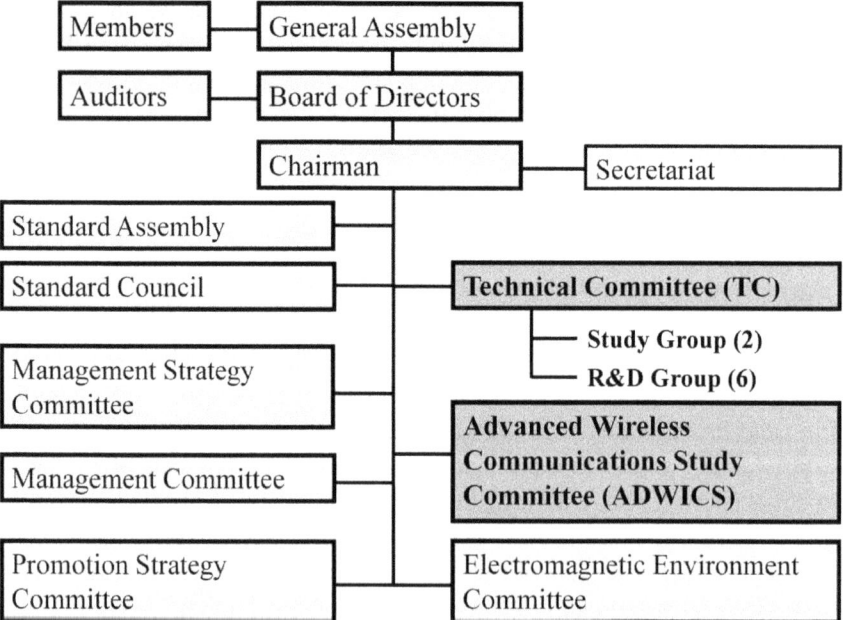

Figure 1 Organization of ARIB

2 Development of the ARIB STD and TR

ARIB, as an SDO, conducts study and R & D relating to the utilization of radio waves from the viewpoint of promoting practical application and dissemination of radio systems. Specifically, draft standards (STDs) and technical reports (TRs) are developed in two phases, i.e. "study" on the trends of demands and technologies, and "R & D" to specify new radio systems. The Technical Committee (TC) and the Advanced Wireless Communications Study Committee (ADWICS) play a fundamental role in this work. Two "study groups" and six "R & D groups" under the TC and five "subcommittees" under the ADWICS are organized in ARIB.

The outcomes from R & D groups and subcommittees are reported as draft to the TC and the ADWICS, respectively, and then various ARIB STDs and TRs are set in the Standard Assembly. Figure 2 shows the procedure for the development of ARIB STDs and TRs.

As of July 3, 2013, ARIB holds 152 STDs and 64 TRs as shown in Table 1. Among them, the major STDs include digital television broadcasting, digital sound broadcasting, terrestrial multimedia broadcasting, mobile communication systems, and Integrated Transport System (ITS).

The ARIB STDs and TRs can be downloaded free from the following URL:

http://www.arib.or.jp/english/html/overview/index.html.

3 Overview of the Technical Committee – Study Groups

3.1 Telecommunication Field

Currently, there is no active study group under the technical committee in the telecommunication field. Please see Chapter 6 for studies on advanced radio technologies including IMT.

Table 1 Number of STD and TR (As of 3July 2013)

Field	STD	TR
Telecommunications	92	23
Broadcasting	60	40
General	0	1
Total	152	64

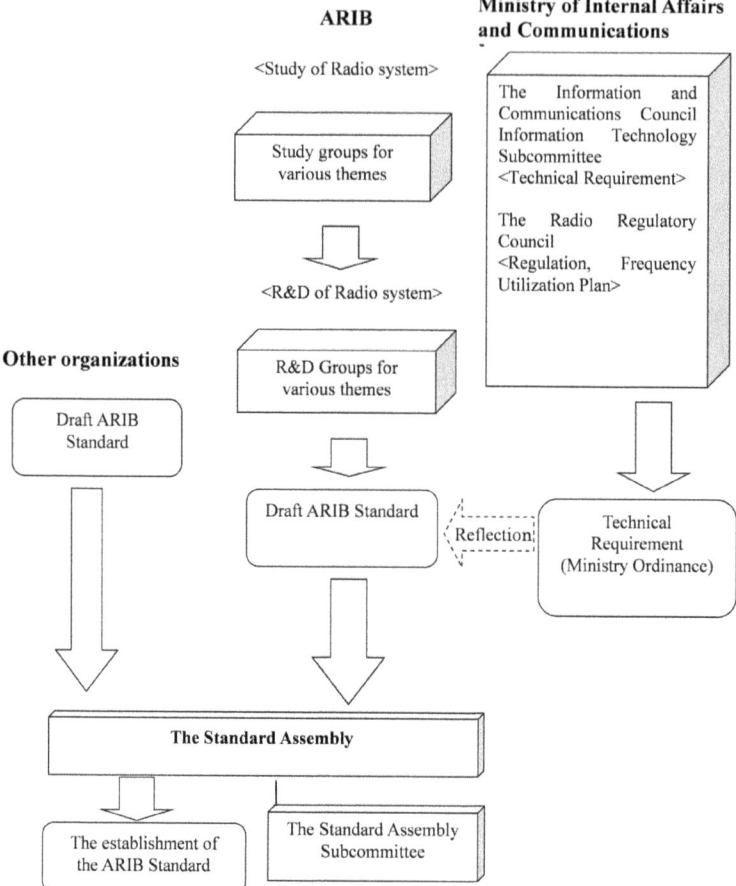

Figure 2 Procedure for development of ARIB's STDs and TRs.

3.2 Broadcasting Field

3.2.1 Study group on the quality evaluation method for broadcasting

The purpose of this group is to study the evaluation method for the quality of video and sound in program content. The major activities are as follows:

- The study of the evaluation method for the quality of video and sound of program content in each production, transmission, and reception field, and
- The support of standardization for the quality evaluation method.

This group is working on the study of the quality evaluation method, the establishment of the guideline, and the production of the test sequence for a new broadcasting system and the related equipment.

It is progressing toward establishing a guideline for maintaining video quality during the real time transmission system of programs and news material over IP networks, as well as monitoring the quality of file-based program material and completed programs. The study extends a quality evaluation of the sound source in digital audio systems.

The group has improved the quality evaluation method for flat panel displays with regard to viewing conditions, such as monitoring characteristics and viewing distance. The results have been given to the International Telecommunications Union-Radio communications Sector (ITU-R).

The production of a new test chart for still images, which corresponds to high resolution (4k and 8k) as well as the wide color range, is also in progress.

3.2.2 Study group on new technology for a next generation broadcasting system

The purpose of this group is to study new broadcasting technology. The major activities include:

- The study of technology for a future 3D television,
- The study of a new service for hybrid broadcast broad band digital television (DTV), and
- The study of a transmission technology for next-generation digital broadcasting.

This group is surveying the trend for future glassless 3D technology, which is expected to be totally different from the present stereoscopic 3D system.

With respect to the new service, the group is studying the capability of Internet-connected TV in order to design a new service. It is also considering the key issues to realize this. Taking into account that the penetration of mobile phones and smart phones with a digital TV function would expand the environment with which to enjoy TV services, viewer expectations regarding such new services would need to be investigated.

For transmission technology, the group is considering a terrestrial broadcasting service with more than 100 M bit/sec transmission rate in the ultra-high frequency (UHF) TV band, along with the utilization of a sub-millimeter and millimeter over the 100 GHz band. The next step would be to design the system image for digital broadcasting in the next generation, followed by identifying the issues to be studied.

4 Overview of the Technical Committee – R & D Groups

4.1 Telecommunication Field

4.1.1 R & D group on broadband wireless communication systems for public use

The 32.5MHz bandwidth (170 − 202.5MHz) within the newly available spectrum by the digitization of terrestrial TV broadcasting has been allocated for the broadband wireless communication systems for public safety.

Since this system consists of a portable base station (BS) and multiple mobile station (MS)s, it can be operated when and where it is needed. It is capable of providing several Mbps transmission data rate within the coverage area of several km from the BS.

One representative use case is high quality video image transmission from disaster areas to the local emergency headquarters. Standardization activity on technical specifications of the system for public use has been carried out in R & D group on broadband wireless communication systems and then ARIB STDT-103 has been developed.

4.1.2 R & D group on wireless lan systems

In response to rapid expansion of the use of wireless LAN systems, beginning with the off-load measures for smart phone traffic, R & D group on wireless LAN systems was established in April 2013 with a view to improve the reliability and advance the wireless LAN systems. The group will research and standardize wireless LAN systems with the cooperation and coordination with related standardization bodies.

4.2 Broadcasting Field

4.2.1 R & D group on digital broadcasting systems

The purpose of this group is to research and standardize the technical specifications, from transmission equipment to receivers in the digital broadcasting system. The major activities are as follows:

- R & D and standardization for the specifications with respect to multiplexing, the coding technique for video, sound, and data, data broadcasting,

copyright protection, and the management of access control, which are common to all systems.
- The R & D and standardization of the digital TV receiver for satellite and terrestrial, the digital sound broadcasting receiver for satellite and terrestrial, and the terrestrial mobile multimedia broadcasting receiver.
- The R & D and standardization of the transmission system for digital TV broadcasting for satellite and terrestrial, the digital sound broadcasting for satellite and terrestrial, and the terrestrial mobile multimedia broadcasting.

This group is a home for the standard of the digital television broadcasting (ISDB) family. ISDB-T for terrestrial TV and ISDB-Tmm for mobile multimedia broadcasting are among them.

STD and TR for the "area-limited ISDB-T" broadcasting system to be operated in the white spaces in the UHFTV band have been developed as ARIB STD-B55 and ARIB TR-B35.

An examination has been conducted to ease the sound level difference between stereo programs and down-converted 5.1-channel surround programs and then ARIB STD B-21 and the relevant ARIB TRs have been amended.

The group also devotes itself to devising specifications for 3D television systems. Some or all of the following points are considered when making various 3D television systems:

- Can 2D receivers display 3D programs?
- Is new video coding technology necessary?
- Can 3D program production be subject to 2D display, and
- Is the additional transmission band necessary?

Considering the necessary transmission bit rate and resolution for each system, as well as the available bit rate in conventional broadcasting systems, the most appropriate system will be selected. In addition to those mentioned earlier, the group makes efforts toward investigating technological trends in domestic and foreign organizations with respect to 3D television systems.

4.2.2 R & D group on program production systems

The purpose of this group is to research and standardize program production systems, editing systems, and transport system in and between broadcasting stations. The major activities are as follows:

- R & D and standardization for the transport system regarding the various auxiliary data,

- R & D and standardization for the interconnection specification between program production equipment,
- R & D and standardization for the measuring method for program production equipment,
- R & D and standardization for the data broadcasting program production system,
- R & D and standardization for the closed-caption program production system, and
- R & D and standardization for the operational technique of program production equipment.

New items being discussed include online exchange for the packaged program and the security of program files.

4.2.3 R & D group on the transmission of television program contribution

The purpose of this group is to research and standardize the transmission system for television program contribution at broadcasting stations. The major activities are as follows:

- R & D and standardization for the digital Field Pickup Unit (FPU) system, the transmitter-to-studio link (TSL) system, the sound contribution system, wireless microphones, and the high-capacity television program contribution in the millimeter frequency band,
- R & D and standardization for digital satellite newsgathering (SNG) systems, and
- R & D and standardization for the digital transmission system for domestic television networks.

In Japan, the frequency for the digital FPU system in the 700 MHz band is scheduled to move to the 1.2 GHz band and the 2.3 GHz band in order to expand the allocation for the mobile cellular system including LTE (Long Term Evolution), so called 3.9G in the 700 MHz band. The work is progressing, taking such domestic frequency re-allocation into account.

4.2.4 R & D group on ultra-high-definition television broadcasting systems

The purpose of this group is to research and standardize program production equipment that realizes a television broadcasting system with more than 1080 scanning lines. The major activities areas follows:

- R & D and standardization for video systems,
- R & D and standardization for sound systems, and
- R & D and standardization for the interface between equipment.

The group has been contributing to the ITU-R in order to devise the ultra-high-definition television (UHDTV) standards.

Considering, the progress in the ITU-R, the draft of the standard for a multichannel sound system that exceeds 5.1 surround channels is being prepared. The interface between the equipment will, based on video format specification, be standardized considering the trend in both the ITU-R and SMPTE.

In response to the establishment of the ITU-R Recommendation BT.2020, the group has researched and standardized the parameter values for UHDTV systems for production and international programme exchange, and then ARIB STD-B56 has been developed, choosing the parameters from the ITU-R Recommendation.

5 Overview of Advanced Wireless Communications Study Committee

Advanced Wireless Communications Study Committee was established in April 2006, through the reorganization of IMT-2000 Study Committee that focused on IMT. The objectives of the Committee is to conduct technical studies and contribute to international standardization in the areas of advanced radio technologies including IMT.

The scope of the Committee has been broadened several times, taking into account the development of mobile communications technologies. Currently, it covers IMT, Broadband Wireless Access (BWA) systems, Mobile commerce, Machine-to-machine communications and other wireless technologies. The Committee consists of five Subcommittees, i.e. Mobile Partnership Subcommittee, Standardization Subcommittee, BWA Subcommittee, Mobile Commerce Subcommittee and Multimedia Mobile Access Communications (MMAC) systems Subcommittee.

5.1 Mobile Partnership Subcommittee

The purpose of the group is to operate 3GPP, 3GPP2 and oneM2M as one of the Project Partner (Owner) and to facilitate ARIB member's activities in the Partnership Projects. The major activities are:

- Participate in 3GPPs/oneM2M as one of Organizational Partners of the Projects,
- Information exchange on 3GPPs/oneM2M activities and members' support in the Partnership Projects,
- Prepare contributions to 3GPPs/oneM2M on national regulatory aspects and transpose 3GPPs/oneM2M specifications to ARIB standards, and
- Information exchange on OMA (Open Mobile Alliance) activities.

Under the subcommittee, there are four 3GPPs support working group (WG)s, 3GPP Meeting Invitation Group (Japanese Friends of 3GPP), oneM2M support WG, and SIG (Special Interest Group)-OMA. The current main activities of 3GPPs support WGs are information exchange on relevant 3GPPs groups as well as reflection of national regulatory matters such as new radio band class into the specifications.

3GPP Meeting Invitation Group plans and coordinates to host 3GPP meetings in Japan. As for oneM2M support WG, the current main task of the group is the refinement of details of the oneM2M Partnership Project through the discussion with other Partners. SIG-OMA conducts information exchanges on OMA activities to facilitate the activities of the members.

5.2 Standardization Subcommittee

The purpose of this group is to conduct technical studies on future IMT (beyond what is studied in 3GPPs and IEEE) and to promote its international standardization through contributions to ITU and other activities. The major activities are the following:

- Study issues related to ITU-R WP5D and contribute to ITU-R WP5D,
- Cooperate and coordinate with related international/domestic bodies on IMT Standardization, and
- Prepare SDO contributions to ITU.

The main activity of this group is to study the ITU-R WP5D issues and contribute to national/international activities related with ITU-R WP5D. Collaboration Group (CG) under the Subcommittee conducts collaboration with external bodies, and its current main activity is collaboration with CCSA (China Communications Standards Association), TTA (Telecommunications Technology Association of Korea) on the issues of WP5D under the cooperation framework among 3 countries' SDO. At the each meeting among 3 SDOs, information and opinions has been exchanged, and the draft joint contributions for WP5D were often developed.

5.3 Broadband Wireless Access (BWA) Subcommittee

The purpose of this group is to conduct technical studies on BWA systems and standardization of their specifications as ARIB standards. The major activities are:

- BWA systems in the 2.5 GHz band of which technical requirements have been studied by Information and Communications Council,
- Information exchange on IEEE, WiMAX Forum and XGP Forum activities, and
- Transpose IEEE/WiMAX Forum/XGP Forum specifications to ARIB standards.

BWA subcommittee consists of four working group (WG)s. They mainly assume a role in producing national standards for Mobile WiMAX, IEEE802.20 and eXtended Global Platform (XGP) referring to relevant international standards.

International Relations WG establishes coordination framework with relevant international standardization bodies.

WiMAX WG develops and maintain ARIB standard for Mobile WiMAX system and to liaise with relevant international standardization bodies.

802.20 WG develops and maintains ARIB standard for IEEE802.20 TDD Wideband and 625k-MC Modes and to liaise with relevant international standardization bodies.

XGP WG develops and maintains ARIB standard for XGP system and to liaise with relevant international standardization bodies.

5.4 Mobile Commerce Subcommittee

The purpose of this group is to promote the development and standardization of Mobile commerce and to contribute to the growth of mobile contents market. The current major activities are the following:

- Study possible methods to incorporate official personal identification into mobile phone,
- Identify issues on the access to electronic government from mobile phone and study their solutions, and
- Study payment applications, using NFC built-in mobile phone.

5.5 MMAC Systems Subcommittee

The purpose of this group is to conduct technical studies on MMAC systems and standardize their specifications as ARIB standards. MMAC systems are portable Wireless Access System which can transmit ultra-high speed, high quality Multimedia Information using the SHF band (3 − 30GHz), the millimeter wave radio band (30 − 300 GHz) and other band. The major activities are the following:

- Study on MMAC systems specifications,
- Information exchange and popularization activities related to MMAC systems, and
- Transpose IEEE specifications to ARIB standards.

MMAC systems subcommittee consists of three working group (WG)s. They mainly assume a role in producing national standards for IEEE802.11 and other original specification.

MMAC802.11 WG conducts study on the high throughput WLAN that are under development in IEEE 802.11, maintenance of ARIB STD-71, and drafting of its revision.

Ultra Wide Band (UWB) WG conducts experimental test of the propagation of UWB system considering real environments, exchange of information regarding the interference mitigation function among Asian country, maintenance of ARIB STD-91, and drafting of its revisions.

Wide Area Sensor Network (WASN) WG conducts study on WASN systems using many simple radio data terminals distributed over a wide area which perform data communications and information processing, study on specifications for WASN systems, and study on the mandatory requirements for government standards.

6 Collaboration with other SDOs

ARIB conducts activities on the liaison, coordination and cooperation with the International Telecommunication Union (ITU) and foreign standards development organizations related to radio systems in the field of telecommunications and broadcasting and the dissemination of the radio systems that was developed as ARIB Standards to foreign countries.

Table 2 shows the cooperation agreement, etc. that currently ARIB has entered into with the foreign standards development organizations, etc.

CCSA: China Communications Standards Association

Table 2 List of cooperation agreement, etc. with ARIB

Organization	Title	Date
TTA	Memorandum for mutual cooperation between ARIB and TTA	April 3, 1996
RAPA	Memorandum for mutual cooperation between ARIB and RAPA	April 3, 1996
3GPP	Third Generation Partnership Project Agreement	December 4, 1998
3GPP2	Third Generation Partnership Project Agreement for 3GPP2	January 27, 1999
CCSA, TTA, TTC	Memorandum of Understanding for mutual cooperation among CCSA, ARIB, TTC and TTA	November 7, 2002
KORA	Memorandum of Understanding between KORA and ARIB	December 12, 2005
SMPTE	Memorandum of Understanding between ARIB and SMPTE	February 14, 2007
ICU	Memorandum of Understanding between ARIB and ICU	June 23, 2009
GISFI	Letter of Intent between GISFI and ARIB	October 15, 2009
ITU, ARIB, CCSA, TTA, TTC	Memorandum of Understanding between ITU, ARIB, CCSA, TTA and TTC	July 6, 2011
ETSI	Co-operation Agreement between ETSI and ARIB	November 1, 2011

ETSI: European Telecommunications Standards Institute
GISFI: Global ICT Standardization Forum for India
ICU: Info communication Services Market Participants Union
ITU: International Telecommunication Union
KORA: Korea Radio Station Management Agency
RAPA: Korea Radio Promotion Association
SMPTE: Society of Motion Picture and Television Engineers
TTA: Telecommunications Technology Association
TTC: Telecommunication Technology Committee

7 Recent Achievement

Recent achievement of ARIB includes the development of the following new ARIB STDs and TRs.

7.1 Telecommunication Field

- 1.2GHz TV Whitespace-Band Land Mobile Radio Station for Specified Radio Microphone (ARIB STD-T112)
- 79GHz Band High-Resolution Rader (ARIB STD-T111)
- DSRC Basic Application Interface (ARIB STD-T110, TR-T22)
- 700MHz Band Intelligent Transport Systems (ARIB STD-T109, TR-T20)
- 920MHz-Band Telemeter, Tele control and Data Transmission Radio Equipment (ARIB STD-T108)
- 920MHz-Band RFID Equipment for Specified Low Power Radio Station (ARIB STD-T107)
- 920MHz-Band RFID Equipment for Premises Radio Station (ARIB STD-T106)
- Wireless MAN-Advanced System (ARIB STD-T105)
- LTE-Advanced System (ARIB STD-T104)
- 200MHz-Band Broadband Wireless Communication Systems between Portable BS and MSs (ARIB STD-T103)

7.2 Broadcasting Field

- UHDTV System Parameter for Program Production (ARIB STD-B56)
- Transmission System for the Area-limited Broadcasting (ARIB STD-B55, TR-B35)
- 4FSK Modulation Push-to-Talk Radio Communication Line for Broadcasting Operation (ARIB STD-B54)
- Receiver for Terrestrial Mobile Multimedia Broadcasting Based on Connected Segment Transmission (ARIB STD-B53, TR-B33 ISDB-Tmm)
- Operational Guidelines for Loudness of Digital Television Programs (ARIB TR-B32)

8 Conclusion

With the rapid advances in digital technologies, radio systems in the field of telecommunications and broadcasting has become a fundamental infrastructure that supports economic activities and social lives as well as creates a new culture.

To make further advancement of radio systems effectively, international cooperation to establish unified international standards for radio systems should be important.

With this point of view, ARIB will continue to conduct activities such as study and R & D on the effective utilization of radio waves in the field of telecommunications and broadcasting, development of standards, as well as liaison, coordination and cooperation with related foreign organizations, which, ARIB hopes, would contribute to enrich the ICT society.

Biography

Dr. Kohei Satoh joined the NTT Laboratories in 1975, and transferred to NTT DoCoMo in 1992. He has been a President and CEO of DoCoMo Communications Laboratories Europe GmbH in Munich, Germany, from November 2000 to May 2002.

In July 2002, he moved to Association of Radio Industries and Businesses (ARIB). He is now a Managing Director of ARIB, and his current job is to promote standardization activities on mobile communication systems for enhancement of IMT-2000 and for IMT-Advanced.

Since he joined ARIB, he has actively participated in ITU-R, 3GPP and other regional and domestic standardization activities for IMT. He is currently vice chairman of APT Wireless Forum and vice chairman of 3GPP Project Coordination Group.

New Approaches for Task Classification about Standardization Skills

Toshiaki Kurokawa

ICES Founder, independent design thinker, toshiakikr@gmail.com

Received: 25 June, 2013; Accepted: 21 August, 2013

Abstract

Japanese experts group has developed a set of skill standards for standardization professionals [1]. However, the approach adopted for classifying the tasks of standardization professionals is a traditional and limited one which does not cover the whole range of activities performed by standardization professionals, and moreover neglected some of the important skills needed for people who are dedicated to standardization. In this paper, the author will try to sketch new and alternative approaches for task classification about standardization skills. One important aspect is the Standardization for Public, the other aspect is the evolutional stage of standards proposed by Ken Krechmer [2].

Keywords: Skill standard, skills for standardization professionals, task classification, standardization for public, evolutional stages of standards.

1 Introduction

Standards and standardization mean various things for various people. However, the importance of standards and standardization is well recognized these days. For example, recent APEC textbook on standards and standardization [6] refers various important cases for standardization and the merits of standardization. The major benefits are described for for-profit corporations so is the motive for recent development of skill standards for standardization professionals developed in Japan supported by Japanese Ministry of Economy, Trade and Industry (METI) [1].

In their recent paper [7], authors of Study Group on Skill Standard for Standardization (SG-SSS) describes the reason for defining the skill set for standardization professionals as follow;

In order to collect appropriate human resources for actual tasks for standardization and to carry out effective education of standardization, it becomes indispensable

- to clarify the tasks required for standardization, and
- to clarify the skills required for the tasks.

The output of their work is the definition of 36 tasks identified for standardization, and skills defined to perform these tasks along with the three levels of skill evaluation criteria.

Task definition is done through the popular classification of de jure standards, forum/consortium standards, de facto/company-product standards and house rules, along with the following steps of standardization management:

- Strategy: strategy planning, tactics planning, founding organization, managing organization,
- Development: developing technology, developing standards, managing organization.
- Implementing: applying standards, acquiring certification.
- Promotion: promotion planning, advertising.

Skill evaluation criteria are categorized either on performance or on capability.

There are 4 aspects for performance evaluation: Responsibility, Experience, Achievement, and Contribution. And the followings are the aspects of capability evaluation: Business comprehension, Communication, Negotiation, Planning, Leadership, Presentation, Technology, and Operation.

The skill levels are the following three levels: Low for trainee, Middle for autonomously acting person, and High for leading figures instructing others as well.

So, the total descriptions are summed up around 1,200 detail description to cover the entire skills for standardization professionals. However, even in the midst of compiling these documents with Japanese experts, there emerged some criticism for this scheme of skill definition.

The major problems lie in the classification of tasks. It should be noted that the task is not classified from the standard/standardization itself but the management of standardization from for-profit corporation.

The following Figure 1 shows the life cycle of standards. Here, we cannot see the task for strategy, implementation and promotion. They are part of the standardization management task for for-profit corporation as shown in Figure 2.

One criticism comes towards this prerequisite of the skill, that is, for the activity for "for-profit corporations." Their target is for business. We will discuss further on this point, but the argument goes that there are non-profit

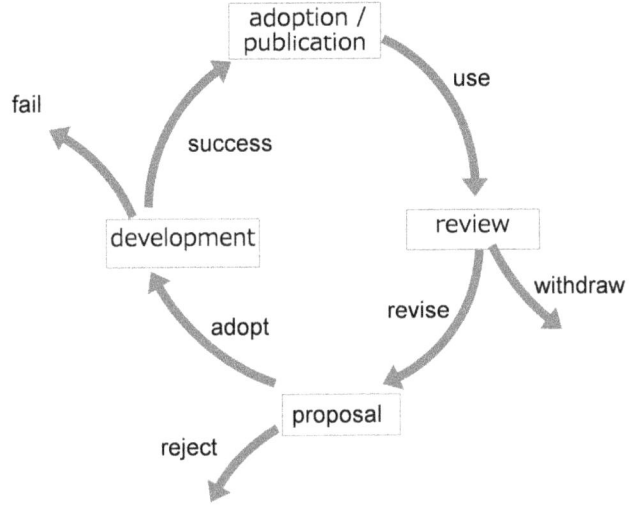

Figure 1 Standards life cycle

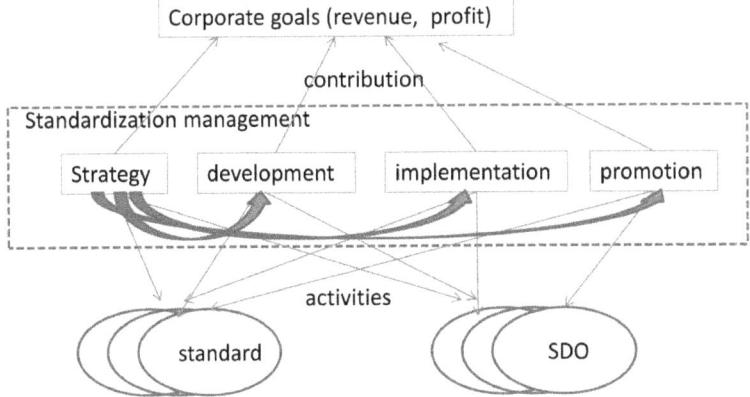

Figure 2 Standardization management tasks for "for-profit corporation"

organizations that are dedicated for standardization, typical examples are ISO, IEC, and ITU, the Standard Development Organizations (SDOs), and the skills for those people in these organization require different skills. Both people may share the same kind of skills, but non-profit organizations' people cannot target for business.

Moreover, there are some aspects for standardization professionals who need to consider the public wellness other than the profit/merits for their company. This aspect for working for the society rather than his/her own organization is one of the great characteristic of activities in standardization, and also one of the most difficult part to be appreciated within the "for-profit organization." We will discuss this aspect further in the next section.

The other problem has come from the fact that the skill set among de jure, forum/consortia, and de facto do not differ much. Major parts of the skill set are common among them, which might be reasonable because nowadays lots of forum standards are converted into the de jure standards, and in a way, they differ only in the decision process and/or certification.

We need some other approaches to tackle this problem to classify tasks for standardization. One suggestion has been given to utilize evolutional model for standards which has been developed by Ken Krechmer [2]. His model is based on the main feature of standard: symbols, measurements, similarity, compatibility, and adaptability. Just like patents, this model describes the differences on standards, or what aspect of products and/or services is standardized, and in terms of standardization, the person in charge should have the skills specific to that kind of standardization.

It is also noted some researchers do not distinguish de jure standardization from forum/consortia standardization. For example, Büthe & Mattli [5] categorize them both in Standardization by Private Agencies without Market Mechanism.

We will discuss the evolutional or feature model for task classification of standardization. Of course, there would be yet other approaches for classifying and defining tasks for standardization.

2 Standardization for Public

In 2006, IEC has published a web book entitled "Standards for business" [3] which has two subtitles: "International Standardization as a Strategic Tool," and "How companies benefit from participation in international standards setting." It is also very interesting when you try "standards for public" in search engines such as Google, you have no exact hit for this inquiry while the "standards for business" will get more than 2.6 million hits.

This typically shows how the standards are considered important in business. And also it shows standards are not discussed much in the context of public wellness. However, this raises a key question that the standards has been important mainly for public reasons from the historical perspective. For example, United States Standards Strategy [4] by ANSI cites John Quincy Adams' statement in 1821 to tell the basic standards (in this case, weights and measures) are "the necessaries of life to every individual of human society. They enter into the economical arrangements and daily concerns of every family."

Recent book on standardization [5] also argues for standardization mostly from economical viewpoint, however, it also cites the issue on public interest raised by European consumer group that the international standardization does not well respond to the consumers' concerns.

2.1 Tasks for Standardization for Public

There are many non-profit organizations engaged in standardization. Typical examples are ISO and IEC that are major international standards development organizations, although in strict sense, they are both in private agencies in contrast to the ITU which is an international governmental organization under United Nation.

People under these organizations are, even though they may be borrowed from for-profit corporations, working standardization from these organizational viewpoint, that is, for public not for profit. Governmental and/or non-profit standardization organizations are also standardizing for public. Even though some of these organizations have the mission to help for-profit corporations in standardization, the performance measurement cannot be the profit.

Also, those standardization professionals within for-profit corporations may also have their mission for standardization bodies, where their expected achievement will be primarily on the standardization activity itself. As they belong to a for-profit corporation, they share the concern of the for-profit corporation but their role in standardization may not directly produce the profit of their employer.

Even though the standardization they are involved is evaluated in the for-profit corporation's strategy for business, the proposed standard itself must have the aspect for the public good. Otherwise, no other people/groups agree with the standardization.

Now the question comes to the point where how much of those tasks for standardization for public differ from tasks for standardization for for-profit corporations. Tasks can be categorized by the strategy/goal related ones and the operational ones which may be derived from the strategy/goals.

Clearly the goals are different for public needs and for for-profit corporation's needs. Then, also, the stakeholders may differ. For example, standardization for public needs to involve users of standards, especially consumer groups, and also needs to work with government and regulation agencies for public wellness. In the case of standardization for private interests, those consumer and government groups are the secondary audience, and this situation has been sometimes criticized.[5]

We will further investigate in two cases: one for those who work in non-profit organization and those who are employed in for-profit corporations but their primary job is for non-profit operation.

2.2 Tasks for Standardization Professionals Working in Non-Profit Organization

Entire tasks for people under non-profit standardization organization can be labelled differently from those working for for-profit corporations. However, most of the tasks for strategy can be applied both for non-profit and for-profit organizations. The Japanese skill standards [1] show clearly the common tasks such as: Strategy planning for standardization, Information collecting/analyzing/evaluating and tactics planning, Supervising, and Liaison establishing. Yet, the details, for example, of Liaison establishing is "lobbying activities with government and standardization organizations are performed to share information and to establish a close liaison-ship with them" are quite different in the non-profit organization. It would be more coordinating activities with other non-profit organization and/or governmental agencies. No lobbying should be necessary.

Major differences may occur in the Development tasks because these Standards Development Organization itself do not produce standards but coordinating SCs and TCs and networking Professionals and Specialists all over the world, so that they can produce standards documents.

Implementing and Promotion tasks also might be quite different. In the case of governmental organization, both tasks can be similar to for-profit corporations, but the non-profit non-government organization may not have the capability of implementation but only for helping others to implement. The objective of promotion would be different from the for-profit organization. In the case of non-profit organization, the promotion could be more general than the specific standards.

2.3 Tasks Migrated in Standardization Professionals Hired in For-Profit Corporations

This is one of the most intrigued and also most important aspect of standardization task itself. In Japan, Skill Standards are defined in some areas, notably in Information Technology (IT) [8] and in Intellectual Property Rights (IPR) [9]. Their skill sets and tasks are simple compared with the tasks for standardization professionals. Also this is more true for those experts working for non-profit standardization organization while hired by for-profit corporations.

IT Skills are for doing some IT projects covering strategy setting and maintenance. IPR Skills are also same kind and for specific for-profit corporations to deal with their own intellectual properties.

In the case of standards, the standard, even if it was drafted within the for-profit corporation for their profit maximization, does not belong to the corporationunless they are de facto standards. Lots of standardization professionals in for-profit corporationsare working for SDOs (Standards Developing Organizations) and managing both for their employers and for standardization communities.

Lots of literatures are devoted for standardization strategies for business, but they do not talk about the standards/standardization for public, perhaps because that is not their interest. Standards and standardization is among the many tools for successful business. It may also because they do not recognize the standards especially de jure standards are essentially for public. Their general attitude is to treat a standard like a patent. If you make the standard, you won it, and you can control the market and your business will prosper, just like you have got a patent registered, and control the market and your business is secured. Unfortunately, this is not true for standards. That is not the way standards are made and accepted, even though there are business aspects around the standards/standardization. Some companies utilize standards/standardization effectively so that their business can be sustainable, yet the standard itself does not belong to the company, that is quite different from the case of intellectual property.

This is also the reason behind the difficulty of evaluating standardization professionals in for-profit corporations, who work for activities of SDOs as board members, conveners, committee chairs and committee members. Their manager may tell the standardization professionals quite frankly, "why I must evaluate you highly, while you are not directly working for our company but for the non-profit organizations." IPR experts may not face this kind of dilemma.

It is really important for these standardization professionals to balance their activities between for-profit and for-public. Sometimes they are lucky enough to enjoy the case where their effort for the standardization for public will promise your employer a good business, however there are cases that many people cannot see why the professional's effort for standardization contribute the company's business, its sales and profit.

SDOs have paid a lots of efforts to validate the economic benefits of standards/standardization. Lots of teaching materials are developed for this purpose, visit ISO repository of teaching materials (http://www.iso.org/iso/home/standards/standards-in-education/education_materials-higher-edu.htm), for example.

While the benefit of standardization is validated, the question would come who should pay the all the burden for standardization. So far, the national government and industry has paid the fee, but consumers group can tell that the consumers and people in general has paid the fee through tax payment and consumption of goods and services provided by industry. The problem is that how we can divide the benefits for each stakeholders and distribute the burden to each sector.

And, moreover, the person in the for-profit corporation who works for standardization community has various tasks that may directly related to the business and also may have no direct relation to business yet indispensable to the standardization organization.

We need more detailed analysis for the skills in this area and to search for a good evaluation system for these skills.

3 Task Classification Based on Evolutional Stages of Standards

One of the key question about the classification of standardization tasks is how much technical skills are needed. While each technical standards need technical skills in each area such as IT, mechanical engineering, chemical engineering and so on, how much technologies are involved in standardization in general, is the big question.

When Japanese group adopted the traditional categorization of de jure, consortia/forum, de facto, in-house standardization, it is questionable if there emerge any technical differences between these categories in technical skill aspects. As you see the outcome of their effort [1], there are no big differences in terms of technology, or technical skills needed.

Of course, there will be classification based on the area of technology: electronics, chemical, civil engineering and so on, however, these are just technologies that each TC (Technical Committee) or SC (Sub Committee) are dealing with. And these technical skills are regarded to belong to its own profession or business, not for standards/standardization per se.

3.1 Krechmer's Evolutional Stage Model

Ken Krechmer has proposed evolutional stage of standards [2] which is a reasonable model for technical classification for standardization tasks. With his model, standards are classified in the following stages: symbols, measurements, similarity, compatibility, and adaptability.

They are summarized in the following table:

Here, the symbol standard is the standard for letters and symbols, which is one of the oldest standard but also active in recent days such as IEC 60417, symbols for use on equipment. In this class, the task for standardization include the IPR related on the registration of these symbols, as well as some knowledge of linguistics, psychology, and ideogram and graphic processing.

The measurement standards are in part of metrology and quite different from symbol standard. Nowadays, the measurement standards are very technical and the Nobel prize has been awarded to the recent work [10]. Standardization skills in this area should be very technical, yet, it can be very political as for the United States customary units for measurements.

The similarity standards are attributed as the outcome of industrial age, most noted as the interchangeable parts such as screws and nuts. Krechmer describes this as follow: "Similarity standards, including process standards, safety standards and quality definitions, define the minimum admissible attributes." This kind of standardization can be possible after the mechanical

Table 1 Krechmer's Evolutional Stage Model

Standards	Ages	Major technologies
Symbols	Hunter Gatherer	Communication
Measurements	Agrarian	Metrology
Similarity	Industrial	Interchangeable (admissibility)
Compatibility	Information	Interface
Adaptability	Post-information	Adaptable interface

production of parts, and the establishment of assembly process of products based on these interchangeable parts. For this kind of standardization, you should have skills related to the production and quality control.

The compatibility standards come next, which handles interface standards such as WiFi, the cellular air interface, the Universal Serial Bus (USB 2.0), and Windows^{TM} Applications Program Interfaces (APIs). Krechmer claims that "Standardization of similarity [...] reduces variation and therefore reduces potential innovation. However, the standardization of compatibility increases variation and innovation." Typical example are those new communication industry based on internet protocol.

As with similarity standards, compatibility standards need technical skills to design the interface, and moreover the skills for setting standards. Even though the similarity standards need technical skills and similarity standards are different from the specification of the products themselves, it is more true for interface standards. Also, interface standards may handle patents issue, in the typical case, it will come with the style of patent pool.

Thus, the standardization skills for compatibility standards may span a lot from the purely technical skills to understand the essence of communication protocol to the legal aspect of patent pool, as well as grouping and business management skills for establishing new industry.

The last category Krechmer proposes is the adaptability standards. This is an adaptable interface standards, targeting multiple interface standards co-exit as long as they are adaptable. Typical example that Krechmer refers is the G3 fax machine protocol. Within this framework of adaptability standard, vendors and users both can enjoy the benefit of innovations while retaining the basic services of communication.

For this adaptability standards, the key technical skill is with etiquettes [11] that will enable each interface to be adaptable with the other standards. On the other hand, the adaptability standards are expanded from compatibility standards, and the skills for compatibility standards will be required in this kind of standards as well.

4 Task Classification for Open and Management Standards

There would be other classification approaches than those in preceding sections for standards/standardization as well. One approach would be Open Standards and Open Standardization. This highlights traditional standard and standardization to be called as Close Standard and Close Standardization.

While the definition of Open Standards and Open Standardization varies from people to people, their key gradients would lie upon their process of making (and/or defining) standards and also for the availability of standards.

Their underlying technology is the Internet and open collaboration is the feature of their process.

The other task classification for standard can come from the so-called Management Standards, or Management Systems Standards, which is quite different from the traditional technical standards where the organization's management process is standardized. ISO describes this as "They provide a model to follow when setting up and operating a management system" [13]. Some people classify the traditional standards as technical standards to highlight the differences from these management standards.

However, those Open Standards and Management Standards are less connected technologies than those Krechmer's evolutional model addresses.

5 Summary

In this paper, we have explored non-traditional classification of standardization tasks. This work has been triggered from the Japanese study on skill standards for standardization professionals [1]. Their study has adopted traditional classification with de jure, forum, de facto, and in-house standards along with the processing stages of Strategy, Development, Implementing and Promotion.

Since this traditional classification for standardization tasks is not enough for handling all the tasks for standardization, we have studied several approaches in this paper. These approaches are not exhaustive, so there would be still other approaches as well. We would like to further explore and discuss those new possibilities.

However, it can be noted that the Standardization for Public should have an impact on standardization in the near future, since the voices for promoting standardization for public has becoming popular in recent ICES 2013 Workshop [12] held in June 2013. These voices are high especially for under development countries where farmers, fishermen and small merchants have the strong concern with the standards but cannot have the luxury to own their standardization professionals but only can rely to the standardization staff within government and public service.

Task classification based on Krechmer's Evolutional Stage Model is also important in the sense that these shed some lights on technologies behind the standardization so that the university level curricula can be organized along this set of technologies.

Also more work can be employed for other approaches such as the Open Standards/Open Standardization and Management Standards. These are rather new to the standards/standardization so we may need more experiences with these approaches.

References

1. Skill standard - Evaluation for skills of human resource required for standardization, Ver. 1.03, 2013–05–31, http://docbox.etsi.org/workshop/2013/201306_ices/presentations/7-papers%20and%20posters/kurokawa%20et%20al%20iieej%20japan%20skill%20standard.pdf 2013.
2. Ken Krechmer, "Balanced Standardization" Journal of IEEJ, Vol.41, No.5, pp.568–574, http://www.y-adagio.com/public/committees/std/serial_sv/st4.pdf.
3. Henk J. de Vries, Standards for business - How companies benefit from participation in international standards setting, IEC Centenary Challenge 2005.
4. ANSI, United States Standards Strategy, Third Edition, ANSI, 2010.
5. Tim Büthe & Walter Mattli, 'THE NEW GLOBAL RULERS – The Privatization of Regulation in the World Economy', Princeton University Press, 2011.
6. APEC Sub-Committee on Standards and Conformance, 'Education Guideline 3: Textbook for Higher Education - Standardization: Fundamentals, Impact, and Business Strategy', APEC, 2010.
7. Toshiaki Kurokawa et al., 'Skill standard -Evaluation for skills of human resource required for standardization', Proceedings of ICES Workshop 2013, 2013.
8. IPA, The Skill Standards for IT Professionals, 2008, http://www.ipa.go.jp/english/humandev/forth_download.html
9. METI, The Skill Standards for IPR Human Resources, 2007, http://www.meti.go.jp/policy/economy/chizai/ipss/(in Japanese).
10. Quantum Information Science and Engineering Program at NIST, http://www.nist.gov/pml/div684/qip.cfm
11. Krechmer, K.: The fundamental nature of standards: technical perspective. *IEEE Communications Magazine*, 38(6), pp. 70–80, 2000. Also available at http.//www.csrstds.com/fundtec.html

12. ICES and WSC Academic Day 2013, held at ETSI, June 2013. http://www.etsi.org/news-events/past-events/658-2013-ices-wsc-conference
13. Management system standards. http://www.iso.org/iso/home/standards/management-standards.htm.

Biography

Toshiaki Kurokawa is an ICES Founder and now self-employed as an independent consultant to manage Design Thinking Research & Education at his home. With his colleagues, he formed a joint Study Group on Skill Standard for Standardization (SG-SSS) under IEEJ and IECEJ. At IEEJ, Mr. Kurokawa is the chairperson of SIG on Education about Standardization. He is also an Affiliated Fellow at Science and Technology Foresight Center, National Institute of Science and Technology Policy, Ministry of Education, Culture, Sports, Science and Technology (NISTRP under MEXT) in Japan. He was a Researcher at Intellectual Property Science Laboratory of Kanazawa Institute of Technology, located in downtown Tokyo before February, 2013, where he was engaged in the project of "Evaluation for skills of human resource required for standardization" sponsored by Ministry of Economy, Trade and Industry (METI) in Japan. Mr. Kurokawa has been engaged in standardization through his engineering/marketing/management career at IBM and SCSK (formerly CSK) Corporation from around 1990 mostly in ICT area. In 2006, he founded International Cooperation on Education about Standardization (ICES, http://www.standards-education.org/) with his friends, and has talked his experience at various places including foreign countries such as Indonesia, Malaysia, and USA. He has published numerous papers on education about standardization, ICT and Design Thinking at such journals as "Science & Technology Trends - Quarterly Review." His research covers other areas such as Requirement Engineering, Software Testing, Ontology and Cloud Computing. He is also working for Kids, Their Future & Design for K-6 children, and Senior-Junior Cooperation Initiative to re-vitalize senior experts.

Internet of Things: Architectural Framework for eHealth Security

David Lake, Rodolfo Milito, Monique Morrow and Rajesh Vargheese

Cisco Systems, {dlake, romilito, mmorrow, rvarghee}@cisco.com

Received: 26 June, 2013; Accepted: 8 October, 2013

Abstract

The Internet of Things (IoT) holds big promises for Health Care, especially in Proactive Personal eHealth. To fulfil the promises major challenges must be overcome, particularly regarding privacy and security. This paper explores these issues, discusses use case scenarios, and advances a secure architecture framework. We close the paper with a discussion of the state of various standard organizations, in the conviction of the critical role they will play in making eHealth bloom.

Keywords: IoT, IoE, M2M, eHealth Security, Standards.

1 Introduction

During the 1990s, as the Internet grew in the publics' awareness, a number of e-terms emerged to capture new forms of personal and business interactions. "email" brought new possibilities for people to communicate rapidly and share experiences; "e-commerce" enabled new ways to conduct business and financial transactions through the Internet. The introduction of eHealth brought the promise to improve health and the health care system by leveraging Information and Communication Technologies (ICT).

The precise meaning of eHealth varies with the source. There is not a single consensus definition. Some benefits of eHealth extend from established telemedicine systems; others are only practical using a machine-to-machine (M2M) model and assume that patients have access to broadband service.

The World Health Organization [WHO] defines e-Health as:

"E-health is the transfer of health resources and health care by electronic means. It encompasses three main areas:

The delivery of health information, for health professionals and health consumers, through the Internet and telecommunications.

Using the power of IT and e-commerce to improve public health services, e.g. through the education and training of health workers.

The use of e-commerce and e-business practices in health systems management.

E-health provides a new method for using health resources - such as information, money, and medicines - and in time should help to improve efficient use of these resources. The Internet also provides a new medium for information dissemination, and for interaction and collaboration among institutions, health professionals, health providers and the public."[1]

In speaking about eHealth today, we need to understand the relevance of Machine-to-Machine Communication [M2M] and the Internet of Things [IoT].

The term "Internet of Things" was originally associated with applications that involve Radio Frequency Identification (RFID). These make use of so called tags, tiny chips with antennae that start to transmit data when they come in contact with an electromagnetic field. They are passive communication devices, in contrast to active devices, that can transmit because they have access to a power source like a battery.

The term "Machine to Machine communication" (M2M), describes devices that are connected to the Internet, using a variety of fixed and wireless networks and communicate with each other and the wider world. They are active communication devices. The term is slightly misleading in that it seems to assume there is no human in the equation, which quite often there is in one way or another; hence we favour the term IoT[1].

The term eHealth is widely used by academic institutions, professional bodies, and standards organizations. From most of the definitions, two items keep appearing and seem to be the important concepts – health and technology. The definitions include use of the Internet or other electronic media to disseminate health related information or services.

It is in the home and assisted-living environments where many applications for eHealth are expected to flourish. Monitoring systems for the elderly or post trauma patients has gained considerable attention. These new systems come

---

[1]Lately the more inclusive term Internet of Everything (IoE) has gained traction.

complete with voice and video options. Automated movement monitoring systems allow the identification of falls and notification of medical personnel without any user intervention. Traditional movement monitoring systems are plagued with false alarms. The combination of voice and video allows for verification and a more appropriate response in the case of an alarm.

Another use case is remote monitoring of patients for blood glucose readings, blood pressure, pulse oximetry, or heart monitoring. In the case of blood pressure monitoring, the readings can provide important information to a physician. Furthermore, measurements taken at home, during daily activities, can potentially be of even greater importance to those taken at a doctor's office, since the readings reflect the patient's condition under normal situations.

eHealth brings special characteristics. The monitoring device's environment is a patient; a living and breathing human being. This changes some of the dynamics of the situation. Human interaction with the device means batteries could be changed, problems could be called in to technical support and possibly be resolved over the phone rather than some type of service call. In most cases, the devices on the patient are mobile not static with regard to location.

The environment for monitoring patients has moved from the hospital healthcare services to a patient's context. M2M/IoT eHealth applications enable remote monitoring of patient health and fitness information, the triggering of alarms when critical conditions are detected, and in some cases the remote control of certain medical treatments or parameters.

The rest of the paper is organized as follows. In section 2 we do a deep dive on the role of IoT in eHealth, followed by an architectural framework in section 3. We deal with Security in section 4. Given the importance of security, we pay attention to its different aspects in various subsections (IoT landscape, endpoint devices, networking, cloud-based services, data storage, enterprise, and federated access). Our final section, before the summary and conclusions, is devoted to standards.

2 Internet of Things and eHealth

IoT brings forth a new phase of the Internet evolution that we can characterize as "the Internet meets the physical world". Today's few billions of endpoints will increase in number by several orders of magnitude. This formidable inflation immediately points to an obvious scalability issue. The number game, and consequent scalability issue, may mask deeper issues, though, including the nature of the endpoints, and the nature of the interactions between the endpoints. We will elaborate on this point later.

While the original Internet connected computers, IoT powered e-Health solutions connect information, people, devices, processes and context to improve outcomes. The intelligent devices that are connected (which were previously passive devices and were not connected) provide a great wealth of information that can be used to make actionable decisions based on algorithms and evidence-based models and can significantly impact how healthcare is delivered and operationalized.

Harnessing the power of Internet of Things for eHealth creates a lot of opportunities to improve outcomes and drive wellness is populations there by reducing the strain points of today's healthcare system. Some of the most promising use cases of connected e-health include preventive health, proactive monitoring, follow-up care and chronic care disease management.

According to BCC research: "Prevention must become a cornerstone of the healthcare system rather than an afterthought. This shift requires a fundamental change in the way individuals perceive and access the system as well as the way care is delivered. The system must support clinical preventive services and community-based wellness approaches at the federal, state, and local levels. With a national culture of wellness, chronic disease and obesity will be better managed and, more importantly, reduced."[2]

The changes, uniqueness, opportunities and complexities of e-health enabled by the power IoT is significant and can be characterised by some of the changes that we anticipate in the ecosystem. They include:

- The number of devices that will come online
- The number of devices that will generate information
- The number of decision making points
- The number of entry points into the system
- The number of types of devices
- The number of types of interactions of devices, applications and processes
- The number of opportunities to leverage the data

Just in the US, the market for preventive healthcare technologies and services by 2014 is projected to increase nearly to $16 billion.

This trend is global in nature. The sensor market is an important component of e-health market. The biosensor market is projected to reach $15 billion by 2015 [3]

According to Cisco's prediction there will be 50 billion devices by the year 2020. While the breakdown for healthcare is unknown at this time, the rate at which innovations in the bio sensor market is progressing and the increasing use of the data to inform decisions, healthcare will see its fair share of devices in the IoT space.

The most disruptive changes, though, will depend on our ability to organize endpoints in systems that operate as coherent units to deliver applications of interest. Let us discuss this point further.

The IoT space of terminal endpoints decomposes into two major classes. The first class includes the current smart phones, tablets, and laptops. While each one is quite an advanced technological piece, including sensors and cameras, we can ignore their internal complexity and regard them as simple points, providing connectivity to the person who owns them. There is a second, emergent class of complex systems, not decomposable into just the set of sensors and actuators that integrates them. We could draw examples from Smart Grid, Connected Vehicle, Smart Cities, Manufacturing Automation, etc. Most relevant for our purpose, though, is to emphasize that many of the eHealth endpoints fall into this category. There are several critical distinctions between the two classes:

The endpoint-Cloud client-server paradigm dominates the first class. It is essentially a scheme of communication. This paradigm induces a vibrant model of apps development: tens of thousands of individual contributors develop downloadable apps for Apple's IoS and the Android environments, for example. An important fraction of those apps relate to eHealth (monitoring vital signs, tracking physical activity, etc.).

The second class of endpoints is even richer, and more complex. The endpoints are clustered, organized in coherent systems. These systems require networking, computation, and storage resources. The diverse use cases may require low latency and support for mobility, and also be geographically distributed [4]. Most importantly, many include not only sensors but also actuators. Closing the sensor-actuator loop often imposes strict latency requirements. Development apps for these complex endpoints require domain expertise that exceeds the capacity of any individual contributor. For eHealth the actuation part of these systems have security implications that we will discuss later.

The second class of devices that are the source for the generation of the information can be further classified into various categories.

Based on how the devices are connected to the patient, the devices can be classified into implantable, wearable, unconnected, or connected on a need basis.

Based on how the device is connected to the network, the devices can be classified into wired, wireless, non-connected.

Based on the data that the device generates, the devices can be classified into real-time continues (e.g., patient monitoring), discrete data sources (oximeter that generates data at regular intervals), and one-time data source (e.g., MRI scanner).

Not all data will be created equally – for example, a personal monitor designed to track a long-term trend in a medical condition or its treatment may only require to send data to a processing element every few hours of days, and a delay of a few seconds or minutes would be immaterial. At an extreme, the total loss of data for an entire measurement period where that period is a very small fraction of the total collection time would be of little consequence.

In contrast, a device that actively monitors a serious, life-threatening condition that requires specific action to be taken with a given time period or where a single-occurrence is of importance would impose tight requirements on the collection and dissemination of the data. In that case, it would not be acceptable to delay or lose a single packet of data.

Based on how the device is used by a single person or a group of people, it can be classified into dedicated, shared within a limited group, or shareable with a wider population.

The challenge of an eHealth/IoT architecture is to support this wide range of device types in a variety of care needs and settings.

3 Proposed Architectural Framework

The explosion of Proactive Personal eHealth, self-management of health conditions, and the collection of data, will radically change the manner by which health-care is delivered and information is collected.

Already, national and transnational organisations such as the European Commission have identified and begun work on projects to address issues around scalability, security, data collection, and interoperability. The paper "eHealth Action Plan 2012–2020 – Innovative Healthcare for the 21st Century" [5] details a number of these areas very well, calling out interoperability of systems, legal and societal barriers to adoption, and detailing how through support of the eHealth Network, the European Commission aims to research and solve these problems [6].

Of particular note in the Action Plan is the proliferation of mobile health and wellbeing as detailed on page 9 of the report:

The growth in the mobile health and wellbeing market has been accompanied by a rapid increase in the number of software applications for mobile devices (or 'apps'). Such applications potentially offer information, diagnostic tools, possibilities to 'self-quantify' as well as new modalities of care. They are blurring the distinction between the traditional provision of clinical care by physicians, and the self-administration of care and wellbeing. Network operators, equipment suppliers, software developers and healthcare professionals are all seeking clarity on the roles they could play in the value chain for mobile health.

Coupled with the number of mobile devices "apps" will be specific-use devices, each capable of different levels of security, traffic generation and protocols, each with requirements that mirror their function.

To visualize the architecture framework for IoT enabled e-health, it is very important to understand the lifecycle of the various entities and their interactions. The life cycle of the device data is critical to understand and can be summarized using six C's. They are

1. Connection: the focus for this function is related to how the device is connected to the ecosystem
2. Collection: the focus of this function is related to how data is collected from the sensor. The data can be pushed our pulled from the sensor.
3. Correlation: the focus of this function is related to mapping the data to a context and does correlation to create meaningful and concise data that can be processed and be used to make decisions.
4. Calculation: the focus of this function is to make a decision based on the data that has been filtered and is processed through an algorithm
5. Conclusion: the focus of this function is to take appropriate actions. The action could be to ignore the event or to escalate.
6. Collaboration: the focus of this function is to enable the collaboration between the patient and the care teams.

Architecture for e-health must consider the needs of each step in this life cycle and must address the effective and efficient execution of each function.

The key to e-health architecture is to support an interoperable ecosystem of different types of devices, applications, and backend systems to enable the free flow information for precise and timely decision-making. The

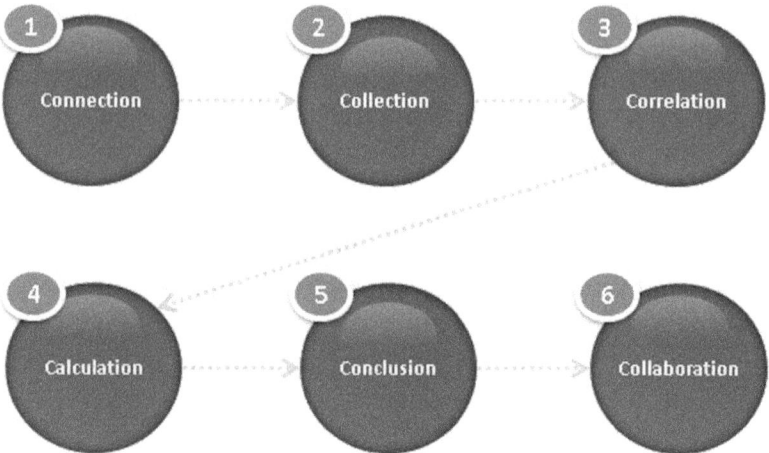

Figure 1 The life cycle of data and processing functions

information service bus enables the communication between the layers and supports multiple protocols.

The device layer must consist of a flexible registry-based model that enables plug-and-play of devices.

Given the number of devices and the information they generate, it is critical that information be filtered. The challenge with filtering is to identify the right information at the right time and eliminate false alarms yet not miss any critical information. The clinical decision support systems are used to process this information to make conclusions and the action that needs to be performed based on the information received from these devices.

The co-relation requires data from multiple systems and hence the architecture must support seamless interoperability between the systems that houses the information. The data includes the real-time data as well as historic data that are stored in the system.

The data flow architecture focuses on the source of the data, the destination the data and path the data. The source of the data is typically the sensor. The data can be either locally cached or is sent to the upstream systems without storing in the sensor. The path taken by the data includes a gateway, which can also cache some of the data and do distributed processing. Intermediate hubs can also store and process the data to filter out or make certain decisions. A distributed rules engine is used to make distributed decisions at the closest point of care. This enables data traffic to be filtered and processed efficiently without having every data being processed by the cloud service.

Internet of Things: Architectural Framework for eHealth Security

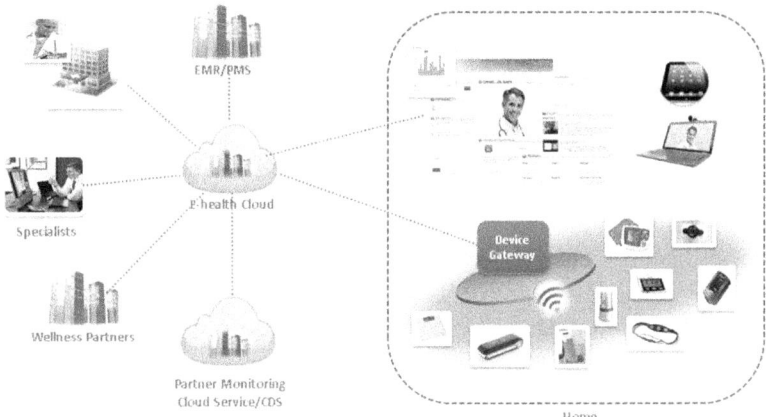

Figure 2 High Level e-Health ecosystem Architecture

The data finally enters the data store in the cloud where it is stored, further processed and archived.

The conclusion could be that a care team member needs to contact the patient to understand further why there is a deviation from the expected readings from the device. This approach is critical to identify problems early in the cycle thereby reducing considerable amount of cost and complexity in dealing with health care issues at the emergency room stage.

Once the conclusion has been made that the care team needs to interact with the patient. Different methods can be used to enable collaboration, which can range from basic text messages to real time video enabled collaboration.

Network architectures must be designed in such a way that these differing, sometimes competing requirements may be met.

The specific use and design of each device will impact the choice of underlying access medium and therefore the level and type of security that can be employed. For example, a device that can be powered from a standard household electrical outlet, have no requirements to be available during a power outage will have less constraints on the network media access layer and encryption algorithm than a worn device which is required to be available 24/7, has constraints on size and therefore on the amount of power that can be associated with network access.

However, as pointed out by the eHealth Action Plan, interoperability or at least commonality in application profiles and data will be key to ensuring an ease of exchange of information between these devices or to the consumers of the information – in order to gain an holistic view of an underlying condition, information is typically gathered from a number of points and by devices with different characteristics, manufacturers, purposes, etc.

Therefore, it is important that a common data-set by employed so that information may be securely consumed by multiple institutions without compromising security.

Looking at the practical implementation of such architecture leads to some requirements that must be met by the components. Generating, moving, and accessing health related information are core activities in any e-Health solution. The associated requirements go beyond those of a traditional data network, and leads naturally to the consideration of a Message Bus, which offers:

- A reliable solution
- Message persistency
- High-performance and scalability
  - Ideally in excess of 100K messages/sec throughput
  - Ability to handle terabytes of messages without performance impact
- Distributed implementation
  - Fault-tolerance with cluster-centric design
- Guaranteed message ordering
- End-End compression support
- Support for online, low-latency communication
- Open interfaces for data connectors

The Apache Kafka distributed messaging system [32] appears as a good fit.

A full assessment of Kafka is beyond the scope of this document. However, in terms of providing a message-bus system appropriate to an e-Health environment, such as distributed system is highly appropriate, offering both peer-to-peer and brokered communications.

For a general discussion of distributed message systems, including Kafka, we refer the reader to the presentation by Max Alexejev [33].

4 Security

4.1 Security in the IoT Landscape

There is a legitimate concern that security vulnerabilities could pose a significant risk to the industry's belief in the ability of M2M/IoT to deliver greater efficiencies and help enterprises optimize the costs of business operations.

Internet of Things: Architectural Framework for eHealth Security 311

More specifically the chances of security breaches increase in direct proportion to the "degree of connectivity" and as more endpoints become connected to the enterprise IT backend systems through public IP networks, the chances of things going wrong, either intentionally or unintentionally, are accelerated.

To evaluate the security architecture for e-health, we break down the architecture into multiple sections and evaluate the security challenges that we have in each of these domains. As depicted in Figure 3, the main domains include endpoint and access, cloud services, partners, and providers.

eHealth applications in an M2M/IoT environment run on a number of components, including sensor devices and actuators, and networking, processing and storage elements. The overall level of security is upper-bounded by the weakest component in this interactive system. Hence, each component, and the overall system must be designed with security in mind.

There are three basic attack vectors, and a corresponding attack surface to each vector. Data is the first attack surface, and the communication channels

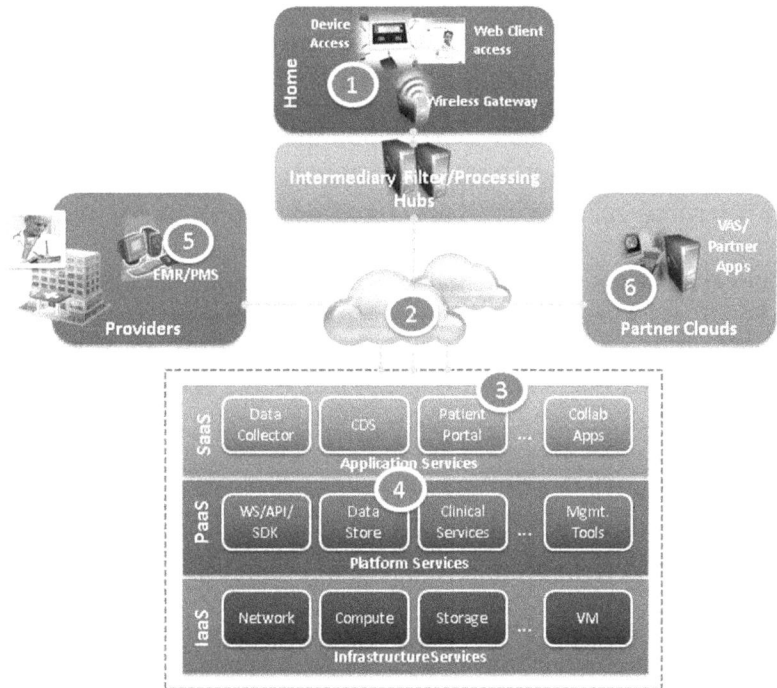

Figure 3 e-health security domain touch points

the second one. M2M/IoT brings forth a third, novel attack surface: physical attacks on or through the medical device.

To deal with the first two issues, one must consider the authentication of the application and/or device, the protection of data, and securing the communication channel, itself. This paper will discuss briefly some of the security issues around device design and use.

A recent editorial [7] welcomes the potential benefits of mobile technology while emphasizing the need to subject the apps to the rigorous standards of evidence-based medicine. Indeed, the smart phone/tablet explosion has spurred a formidable app development activity. A thriving community of tens of thousands of Apple IoS and Android developers has generated a number of interesting apps, a sizable and rich subset of them in the eHealth space. We share with the author the appreciation for the aggregate creativity of this community, which has accelerated bringing apps to the market. We also heartily agree with the need for a strict validation program along the lines suggested by the editorial.

From the security viewpoint we need to go even further, as exemplified by the potential of inflicting physical damage through the compromising of an insulin pump [8]. While developing techniques to prevent hacking of medical devices is beyond the scope of this paper we suggest here a potentially fruitful line of research: a precise definition of the expected interactions for the specific device with the external world, and a clear baseline of the expected behaviour could be the basis for building a tampering-resistant device.

It is important to note that the term "securing the communication channel" is quite broad and should be about ensuring the confidentiality, integrity, and reliability of data sent over telecommunication networks in a connected ecosystem. Security in the M2M/IoT model is not only about ensuring the proper access to the right entities at the right time but also about creating a secure architecture.

The next section will take the reference architecture and a simple version of an eHealth use case to examine the first two aforementioned threat vectors and recommended mitigations.

Consider a patient with a blood pressure monitoring device, which takes a blood pressure reading every 15 minutes and the device itself or another local device, which has a collector function stores the readings. Once a day the device or collector and the medical facility's application server communicate with each other to transfer the reading to the server. The collector function could be in the M2M gateway. The readings may be summarized or some other data manipulation technique performed, and then the medical staff reviews the results.

This simple example illustrates several areas for security to address.

On the device and collector:

- Secure Boot of the device for platform integrity check and boot loader authenticity
- Secure Storage of the secret keys. The storage should be physically tamper resistant and access control protected
- Secure Storage of the data
- Device identification must be a unique identifier within the eHealth context

For the communication channel:

- Mutual Authentication between the eHealth device and the application server and/or the network
- Data Integrity to protect the data from any alteration during the communication session
- Data Confidentiality uses encryption and decryption of data between the secure device and the application server and/or network during data exchanges

In the Ecosystem:

- Key Management of the secret keys in the eHealth
- Cryptographic Support of cryptographic protocols, such as AES and optionally PKI

There are several things to consider in this example.

Does the M2M monitoring device and collector have proper security characteristics?

For the infrastructure components, M2M communication starts with connected devices. Devices have a certain level of processing power that is then used for monitoring or reporting on specific events or conditions. There is a possibility of the firmware being compromised on the device itself. Device OEMs normally rely on their manufacturing partners to develop and configure these elements. These OEMs have to ensure that their partners adhere to strict security policies themselves. A device that goes rogue or does not perform as expected under certain conditions is equally dangerous (if not more) as a device that is open to easy access and manipulation by unauthorized personnel.

Is the local environment considered secure? If not, then all communications from the device should be protected.

The environment in which the monitoring device operates is very important. If the device is only used in a home, or a medical facility, then one might be able to assume that is a trusted environment. In a trusted environment, the communication between the device and the gateway device may not need to be encrypted. However, in any environment that is not considered trusted, then the communication needs to be encrypted for data conditionality.

Currently a prominent insulin pump vendor provides wireless connectivity between their insulin pump and an USB device plugged into the patient's PC. One might consider this a safe and secure environment. However, the communication protocol is proprietary and not secured. Recently a researcher demonstrated that this insulin pump could be hacked into and controlled if one is in the close proximity. This allowed the hacker to instruct the pump to perform all manner of commands, even dispensing the complete reservoir of insulin to the user without the user's knowledge. This could have fatal consequences.

The second issue and the biggest challenge from the device side is that a lot of M2M/IoT devices do not have enough capability to do the encryption on the device. This is a currently a big issue in M2M/IoT. Many current devices and sensors have a small amount of on-board memory and a microprocessor that is simply unable to handle standard security protocols (such as AES 128 and others). Designs are improving on that front, and there are products coming that are capable of 16-bit processing. The other issue on the device side is regarding the amount of battery power it takes to do all the algorithmic computations for data encryption. This is especially relevant in cases where the endpoints communicate directly with the enterprise IT backend systems without having a local aggregation unit. Some remote deployments are expected to run for several years on battery power without any human intervention (to replace the battery for example) and computationally intensive processes could put a significant strain on the batteries.

The algorithms require a minimal amount of power, occupy a very small footprint in memory and handle computational transactions extremely fast. With these types of advancements it may make it easier to meet the security requirements for integrity and confidentiality.

What protocol does the monitoring device support for communication?

The communication protocol that a device uses is not always IP. The Continua Health Alliance supports ZigBee as the preferred Personal Area Network (PAN) protocol for devices. Other supported protocols by the industry are Z-Wave, ANT and others. Therefore, there is a need for an additional device acting as a protocol gateway between ZigBee and IP networks. This device is often used for the collection of data, and is usually connected to main power. The protocol situation is similar to the situation 20 years ago with network protocols. There is work by various groups to help promote the use of IP in devices.

The wireless network provides some degree of security. Unfortunately, the current scheme of encryption over GSM/GPRS networks is not totally secure, although it does take a high degree of technical sophistication to break in these ciphers. A5 encryption has been considered broken for some time. Certainly there are significantly more security mechanisms in 3G and 4G, but currently that is a very small percent of the M2M/IoT market.

A rogue base station software, such as OpenBTS and OpenBSC can be used to launch a man-in-the-middle (MITM) attack. It has been demonstrated that with the use of a rogue base station and a patched cell phone, it is possible to get into a vendor's private network. The cell phone builds a bridge to the vendor's network.

One method for added security is to use an encryption mechanism that is layered - 128-bit AES, and then the A-5 encryption on the GPRS channel. The perpetrator would not be able to break the AES encryption. However, most M2M/IoT device manufacturers have not implemented AES, as the present generation of sensor type of devices that are not that powerful from a security perspective.

4.2 Endpoint Devices Security

The management of long-term and chronic conditions will therefore require a number of worn and/or embedded devices, constrained by their specific purposes in their communication and information exchange.

As discussed in the book "Body Sensor Networks" "Body Sensor Networks" [9] network and media access systems will differ between each of these, as will their security capabilities and needs.

It is envisaged that a common device will act as the collector and orchestrator of this information operating at the heart of an autonomous, but connected system. Rather than imposing a single architecture and security scheme on every device, it makes more sense to choose security attributes that are appropriate to each device, wrapping these to a central policy from the collector as part of an autonomic [10] computing domain.

Assuming that identity of the end device can be managed – many of these wearable devices have identity as part of their composition – managing the identity of the collection device and association with the sensors will be important. Establishing a trust relationship between the collector and the service consumer is vital – technologies such as the Trusted Platform Module[11] as used in compute systems the world over can be embedded in these collection agents.

It is likely that a number of parallel collection agents may be required – for example, for two family members with different conditions or a single person with multiple information sources or where hard separation between data sets is required.

It is suggested that with enhancements to the home-gateway function at the end of most cable, DSL or other Internet connections, the use of virtual machines running in "slices" could provide the ability for a device manufacturer to deploy software images that provide access to use-specific devices. This has the benefit of moving the intensive, power-draining compute functions away from the sensor. Device specific security algorithms and protocols can be employed on the sensor, with translation to systems more appropriate for wide-area network connectivity on the mediating gateway.

Many home-gateway devices also support multiple RF connections – Wi-Fi, Bluetooth, cellular back-up services – run ARM or x86 processors and can support applications today. With the commoditization and rapid price decreases in home-gateway devices, providing this functionality at the edge becomes economically attractive and architecturally beneficial to eHealth.

An example of the MyHeart eHealth architecture in which many of these aspects of sensor usage, security and multiple applications has been piloted is discussed in the referenced paper [12].

According to NIST, the threat profile for handheld devices is a superset of the profile for desktop computers[13]due to the size, portability and availability of wireless interfaces and associated services. The security threats to mobile devices include, but not limited to theft, unauthorized access, malware, spam and electronic tracking. Malware attacks can result in spoofing, data interception, data theft, backdoor access, unauthorized network access, service abuse, and impact the availability and integrity of the data.

The architecture should enable controls and policies that can not only prevent many of these attacks, but also limit the damage in case of a breach. Policies that enforce rules such as authentication, strong passwords, and password changes at periodic interval, disabling services that are not required are critical in ensuring the security.

The use of prevention and detection software in the endpoint devices is critical. These include Data encryption capabilities, firewall, antivirus, intrusion detection, anti-spam, remote diagnostic and auditing software.

If a device is stolen the ability to disable services, lock, wipe out sensitive data is important to limit the damage and is typically performed by MDM solutions, which are important components of the mobile architecture.

The gateway devices can be either hardware or software-based. With the advent of mobile devices, the gateway software is able to use many of the key features that are available in the mobile devices to transfer the device data from the medical device to the final destination. This also provides mobility and a single multipurpose device for the user as compared to a hardware-based gateway.

4.3 Network Security

The e-health network security architecture involves multiple layers of prevention, detection and response controls as the network spans through different types of networks. These include wireless, wired, enterprise, private and public networks.

The mobile security reference architecture published by the department of Homeland security [14] calls out that the devices that use Wi-Fi and cellular network communications are more accessible and exposed than hardwired devices.

The wireless network architecture must consider the protection from various network based threats such as data interception over air, data interception over the network, manipulation of data in transit, connection to untrusted service, jamming [15] and flooding.

The architecture must enable the tuning of quality of service, which can vary based on the devices and the functionality that is used. For example a pulse oximeter generates text data, which is largely transparent to the delays in the network. In contrast stethoscope audio streaming can be extremely sensitive to network delay.

4.4 Cloud-based Application Access Security

e-health catering to multiple actors and patients constitutes one of the most important and largest segments of Health Care. With the patient base being used to accessing consumer-focused applications from anywhere, cloud models that support anywhere and seamless access are important to ensure user adoption of e-health. The architecture must support different models of on boarding, which includes managed-care as well as self-service based models. This requires integration with multiple systems to enable seamless flow of information that is required to on board the device. All these features opens

318 David Lake et al.

Figure 4 e-health security building blocks stack

up security challenges and the architecture must provide protection against web-based threats such as phishing, drive by downloads, exploitation of vulnerable browsers to get access to applications and data.

Authentication, Authorization, and Accounting (AAA)[16] are basic pillars of secure mechanisms to enable secure access to resources in web applications. Secure policies play a key role in making the access control methods effective. For example, a hacker using dictionary-based attempts could access a system featuring AAA capabilities if a user uses a weak password. The architecture must enforce not only secure controls but also limit failed attempts to access the system using lockout schemes.

The security building blocks can be stacked in multiple layers from the physical security aspects, internal application and data protection to the secure interface access controls. The building blocks for application security are shown below.

Open Web Application Security Project (OWASP) publishes the top ten web security flaws at its website. The 2013 list includes the following [17]:

1. A1 Injection
2. A2 Broken Authentication and Session Management
3. A3 Cross-Site Scripting (XSS)
4. A4 Insecure Direct Object References

5. A5 Security Misconfiguration
6. A6 Sensitive Data Exposure
7. A7 Missing Function Level Access Control
8. A8 Cross-Site Request Forgery (CSRF)
9. A9 Using Components with Known Vulnerabilities
10. A10 Unvalidated Redirects and Forwards

While the overall architecture can protect from security threats, it is extremely critical for web application to ensure above-mentioned security vulnerabilities does not exist. A proper vulnerability patching and software upgrade policy must be adhered to ensure that security risks are mitigated on a on going basis.

4.5 Data Storage Security

The architecture that enables security for data at rest must take an expanded view outside of securing the physical storage, since there are multiple dependencies that can result in weak points, which can be used as an entry point to access the data. A compromised application can be used to access the data, or a backup disk can be used to get access to data. The key considerations must include data encryptions (at application level, file level, disk level), access rights (physical, application, user), context based access and alternate path access (backup data disk)

The federal information processing standards publication 140–2 [18] lists the security requirements that need to be satisfied by a cryptographic module utilized within a security system protecting sensitive information and defines four qualitative levels of security.

Level 1: Lowest level of security and allows the software and firmware components of a cryptographic module to be executed on a general purpose computing system.

Level 2: Enhances the physical security mechanisms by adding the requirement for tamper evidence. This level requires role-based authentication and authorization of an operator to assume specific role and functions.

Level 3: Enhances security to prevent intruder from gaining access to critical security parameters using an identity based authentication mechanisms.

Level 4: Highest level of security and provides a complete envelope of protection around the cryptographic module with the intent of detecting and responding to unauthorized attempts to physical access.

4.6 Enterprise Application Access Security

To enable applications to be accessed from anywhere, more and more e-health applications are hosted on the cloud. This eliminates boundaries for access, at the same time creates security challenges. However many enterprise applications exists within the enterprise and is used to store sensitive information. The sensitive personal health information data storage must be clearly separated from the external web access.

The architecture can involve multiple considerations including DMZ with double firewall protection, reverse proxy to separate the access boundaries for enterprise applications and external applications.

Public Web servers enforce access to content using authentication and authorization schemes. These include basic authentication, Digest authentication, SSL/TLS based server and client authentication.

NIST guidelines for securing public web servers calls out some of the weakness of authentication schemes including SSL/TLS: "Several limitations are inherent with SSL/TLS. Packets are encrypted at the TCP layer, so IP layer information is not encrypted. Although this protects the Web data being transmitted, a person monitoring an SSL/TLS session can determine both the sender and receiver via the unencrypted IP address information. In addition, SSL/TLS only protects data while it is being transmitted, not when it is stored at either endpoint. Thus, the data is still vulnerable while in storage (e.g., a credit card database) unless additional safeguards are taken at the endpoints."[19] Hence, the architecture for e-health must account for multilayer security and follow the path of the data and ensure data is secured while stored, in transit and when handoffs happen.

4.7 Federated Secure Access to Partner Cloud Services

E-health applications use services and capabilities from different sources. This requires secure and seamless access to services provided by other partner clouds. Federated identity allows identities to be shared securely between applications both within and across organizational boundaries.

Protocols such as Security Assertion Markup Language[20](SAML) and WS-Federation, OAuth and OpenID Connect are commonly used standards for identity federation.

The architecture must support a federated identity management model using standard protocols to enable secure and seamless single sign-on into these services. The access to the services is commonly done through web services, and hence the architecture must consider the security aspects for public exposed web services.

5 Implications to Standards

The standards landscape for eHealth-M2M-IoT and security is rather nascent. It includes IEEE for wireless, ZigBee Alliance, ITU-T for M2M e-Health Service Layer, Continua Alliance for Use Case profiles and best practices, NIST[21], as well as diverse government initiatives.

The Focus Group on the M2M service layer (FG M2M) was created in TSAG meeting (January 2012) and started in April 2012 [22]. It is expected to conclude its work in December 2013.

M2M (Machine to machine communication) covers very wide area and several standardization activities have already commenced its study in SG13, SG16 and oneM2M. In order to avoid duplicate works with them, FG M2M will focus initially on services and applications for e-health. The specific tasks of the Focus Group are to:

- Perform a "gap analysis" for vertical market M2M service layer needs, initially focusing on applications and services for the health-care market.
- Identify a minimum common set of M2M service layer requirements and capabilities, initially focusing on e-health applications and services.
- Study whether existing APIs and protocols satisfy the above requirements and capabilities to support a common M2M service layer between M2M applications and telecom networks.
- Draft technical reports describing and addressing the gaps and identifying future standardization work for ITU-T in the field of the M2M service layer.
- Support global harmonization and consolidation by inputting final deliverables to the parent Study Group and other relevant Study Groups as appropriate.
- Develop a living list of SDOs, forums and consortia dealing with M2M service layer APIs and protocols, including information concerning their activities and documents in the context of a common M2M service layer platform.

The following figure depicts the Reference Architecture used for the Continua Alliance:

Continua Alliance [23] is focused on establishing industry standards and security for connected health technologies such as smart phones, gateways

Continua E2E Reference Topology

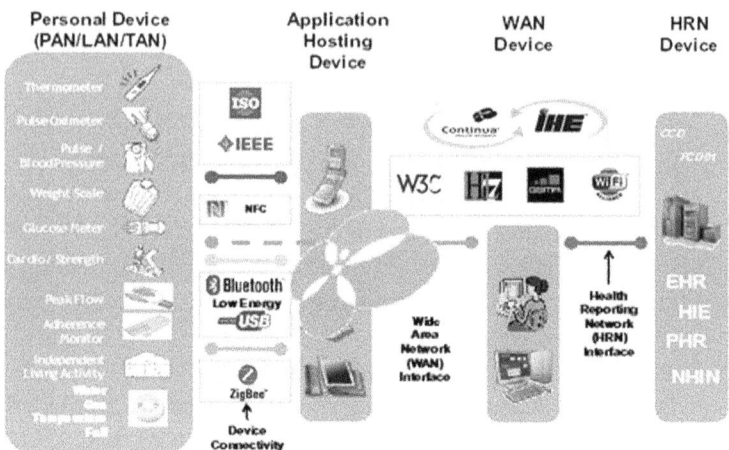

Figure 5 Continua E2E Reference Topology

and remote monitoring devices. Its activities include a certification and brand support program, events and collaborations to support technology and clinical innovation, as well as outreach to employers, payers, governments and care providers.

Multiple Standards exist in healthcare that is used in e-health applications and interactions between humans, devices, processes and applications. These standards can be classified in multiple classes including data standards, message standards, document standards, process standards. They can be syntax based, semantics based, relationship based, purpose based and classification based.

Some of the common data standards include

- ICD (International Classification of Disease) - International standard codes for diagnoses
- CPT (Current Procedural Terminology) - Standard for coding medical procedures
- LOINC (Logical Observation Identifiers Names and Codes) - Standard for Laboratory and clinical observations
- SNOMED CT (Systematized Nomenclature Of Medicine) - Hierarchical Healthcare Terminology
- NDC (National Drug Codes) - FDA's numbering system for medications

The International Classification of Diseases (ICD) is the standard diagnostic tool for epidemiology, health management and clinical purposes. This includes the analysis of the general health situation of population groups. It is used to monitor the incidence and prevalence of diseases and other health problems. ICD codes exist for diseases, signs and symptoms, abnormal findings, complaints, social circumstances, and external causes of injury or diseases. US Healthcare system is in the process of converting from ICD-9 to ICD-10.

IEEE 11073-20601-2008 [24] Standard addresses the need for an openly defined, independent standard for converting the information profile of personal health devices into an interoperable transmission format so the information can be exchanged to and from personal Tele-health devices and compute engines.

HL7 [25](Health Level System 7) provides a framework (and related standards) for the exchange, integration, sharing, and retrieval of electronic health information.

HL7 CDA [26] - Clinical Document Architecture is a XML based markup standard intended to specify the encoding, structure and semantics of clinical documents for exchange. It is based on the HL7 Reference Information Model (RIM) and the HL7 Version 3 Data Types. The purpose is enable exchange of clinical information. It can include multimedia content.

The CCR [27] document standard is used to allow timely and focused transmission of information to other health professionals involved in the patient's care. The CCR data set contains a summary of the patient's health status including problems, medications, allergies, and basic information about health insurance, care documentation, and the patient's care plan

The Continuity of Care Document [28](CCD) is an HL7 CDA implementation of the Continuity of Care Record (CCR). A CCR can be converted to CCD, but not vice versa.

DICOM [29] (Digital Imaging and Communications in Medicine) is a standard for handling, storing, printing, and transmitting information in medical imaging. It includes a file format definition and a network communications protocol. DICOM files can be exchanged between two entities that are capable of receiving image and patient data in DICOM format.

IHE [30] (Integrating the Health Enterprise) is an initiative by healthcare professionals and industry to improve the way computer systems in healthcare share information. IHE defines Integration Profiles, which describe a clinical information need or workflow scenario and document how to use established standards to accomplish it.

Cross-Enterprise Document Sharing [31](XDS) is focused on providing a standards-based specification for managing the sharing of documents between any healthcare enterprise, ranging from a private physician office to a clinic to an acute care in-patient facility and personal health record systems. This is managed through federated document repositories and a document registry to create a longitudinal record of information about a patient within a given clinical affinity domain.

Related to the standards landscape for eHealth-M2M-IoT Security are the various country regulations that define policies for security and privacy specific to health. It is important to monitor these regulations when developing and implementing architecture pertinent to patient safety.

6 Summary and Conclusions

The coming of age of eHealth is intrinsically linked to the successful deployment of a secure and privacy-preserving M2M/IoT infrastructure. The authors have proposed an architecture and framework that support the development and providing of solutions. The authors have further identified core standards and industry bodies where eHealth-M2M-IoT standardization is in progress. While comprehensive, the list is not exhaustive. In closing, we emphasize that security and privacy for eHealth in the emerging IoT landscape offers serious challenges as well as exciting opportunities to the industry.

References

1. http://www.who.int/trade/glossary/story021/en/
2. http://www.bccresearch.com/market-research/healthcare/preventive-healthcare-technologies-hlc070a.html
3. http://www.prweb.com/releases/biosensors/medical_biosensors/prweb8067456.htm
4. http://conferences.sigcomm.org/sigcomm/2012/paper/mcc/p13.pdf
5. "Health Action Plan 2012-2020: Innovative Healthcare for the 21st Century" - https://ec.europa.eu/digital-agenda/en/news/ehealth-action-plan-2012-2020-innovative-healthcare-21st-century
6. http://ec.europa.eu/health/ehealth/policy/network/index_en.htm
7. C. Perera, "The Evolution of e-Health – Mobile Technology and mHealth", http://articles.journalmtm.com/1.1.1-2%20Perera.pdf
8. http://www.foxnews.com/tech/2011/08/04/insulin-pumps-vulnerable-to-hacking/

9. "Body Sensor Networks"–Yang, Guang–Zhong (Ed) ISBN 978-1-84628-484-7
10. "The Vision of Autonomic Computing", http://ieeexplore.ieee.org/stamp/stamp,jsp?to=&arnumber=1160055
11. Trusted Computing Group – http://www.trustedcomputinggroup.org/
12. "An Architecture for Secure e-Health Systems" - http://www.tsb.upv.es/eventos/workshophealthcare/documentos/C2-2.pdf
13. W. Jansen, K. Scarfone, Guidelines on Cell phone and PDA security, NIST 800-124
14. https://cio.gov/wp-content/uploads/downloads/2013/05/Mobile-Security-Reference-Architecture.pdf
15. http://en.wikipedia.org/wiki/Wireless_signal_jammer
16. http://en.wikipedia.org/wiki/AAA_protocol
17. https://www.owasp.org/index.php/Top_10_2013-Top_10
18. FIPS 140-2 http://csrc.nist.gov/publications/fips/fips140-2/fips1402.pdf
19. NIST Publication 800-44: Guidelines for securing public web servers
20. http://en.wikipedia.org/wiki/Security_Assertion_Markup_Language
21. http://csrc.nist.gov/publications/nistpubs/800-66-Rev1/SP-800-66-Revision1.pdf
22. http://www.itu.int/en/ITU-T/focusgroups/m2m/Pages/default.aspx
23. http://www.continuaalliance.org/
24. http://en.wikipedia.org/wiki/ISO/IEEE_11073_Personal_Health_Data_%28PHD%29_Standards
25. http://www.hl7.org/implement/standards/index.cfm?ref=nav
26. http://www.hl7.org/implement/standards/product_brief.cfm?product_id=7
27. http://en.wikipedia.org/wiki/Continuity_of_Care_Record
28. http://en.wikipedia.org/wiki/Continuity_of_Care_Document
29. http://medical.nema.org/standard.html
30. http://www.ihe.net/
31. http://wwww.ihe.net
32. http://kafka.apache.org/index.html
33. http://www.slideshare.net/MaxAlexejev/modern-distributed-messaging-and-rpc

Biographies

David Lake is a consulting engineer in the R&D Group at Cisco. He has more than 20 years of network design and deployment experience, ranging from X.25 and SNA, through the era of multiprotocol routing to IP. He has extensive experience in transporting rich-media technologies across complex enterprise and service provider networks. He is an editor and contributor to the Management and Orchestration (MANO) Working Group with ETSI's Network Function Virtualistion group.

Rodolfo Milito, PhD
Senior Technical LeaderENG Labs
Cisco Systems, Inc Rodolfo Milito, a senior Technical Leader with the ENG Labs of Cisco Systems, is currently engaged in IoT, Big Data and Analytics, and working on a distributed compute, storage, and network platform from the edge to the core of the network nicknamed "Fog Computing".

Rodolfo got his PhD in EE (Control Systems) from UIUC, joined Bell Labs in Holmdel in 1985, and AT&T Labs after the 1996 Lucent spin-off. In 1999 he moved to XStream Logic, a startup in the Silicon Valley, later co-founded ConSentry Networks, and joined Cisco in 2008. Rodolfo's career has straddled research and development in the areas of network design, processor architectures, performance characterization, adaptive control, and algorithms aimed at improving the performance and securing communication networks (load distribution, routing, resource sharing, overload control, and malware detection). He has published extensively, and holds 11 US patents.

Monique Jeanne Morrow CTO Cisco Services Cisco Systems, Inc. Email: mmorrow@cisco.com

Summary

Monique Morrow holds the title of CTO Cisco Services. Ms. Morrow's focus is in developing strategic technology and business architectures for Cisco customers and partners.

With over 13 years at Cisco, Monique has made significant contributions in a wide range of roles, from Customer Advocacy to Corporate Consulting Engineering. With particular emphasis on the Service Provider segment, her experience includes roles in the field (Asia-Pacific) where she undertook the goal of building a strong technology team, as well as identifying and grooming a successor to assure a smooth transition and continued excellence.

Monique has consistently shown her talent for forward thinking and risk taking in exploring market opportunities for Cisco. She was an early visionary in the realm of MPLS as a technology service enabler, and she was one of the leaders in developing new business opportunities for Cisco in the Service Provider segment, SP NGN.

Monique holds 3 patents, and has an additional nine patent submissions filed with US Patent Office.

Ms. Morrow is the co-author of several books, and has authored numerous articles. She also maintains several technology blogs, and is a major contributor to Cisco's Technology Radar, having achieved Gold Medalist Hall of Fame status for her contributions.

Monique is also very active in industry associations. She is a new member of the Strategic Advisory Board for the School of Computer Science at North Carolina State University.

Monique is particularly passionate about Girls in ICT and has been active at the ITU on this topic - presenting at the EU Parliament in April of 2013 as an advocate for Cisco.

Within the Office of the CTO, first as an individual contributor, and now as CTO, she has built a strong leadership team, and she continues to drive Cisco's globalization and country strategies.

Rajesh Vargheese is the CTO for Cisco Healthcare solutions business unit at Cisco. Rajesh leads the technology team that designs, develops and deploys healthcare solutions and leads the effort in defining the strategy, roadmap and architectures for Cisco Healthcare solutions. Rajesh has 18 years of experience in thought leadership, defining architectures, product and solution development, system integration, marketing and messaging, deploying and problem solving in collaboration, video, cloud, real time applications, big data analytics and security for vertical domains. Rajesh currently focuses on the healthcare vertical solutions and brings together his broad experience in collaboration, infrastructure, workflows, clinical system integrations and cloud architectures in developing and shaping the Cisco Healthcare Solutions. Rajesh works with C levels and IT leadership to understand and provide innovative solutions to customer challenges.

Realization of Service-Orientation Paradigm in Network Architectures

Rahamatullah Khondoker[1], Abbas Siddiqui[2], Paul Müller[2] and Kpatcha Bayarou[1]

[1] Fraunhofer SIT, Rheinstr. 75, Darmstadt, Germany
[2] Integrated Communication Systems, University of Kaiserslautern, Germany

Received: 24 May, 2013; Accepted: 8 October, 2013

Abstract

The implementation of communication protocols in the current Internet architecture is tightly-coupled which hinders the evolution of the Internet. This article describes how the principles of Service Oriented Architecture (SOA) can be employed to develop a flexible network architecture. The prototype of the concept has been developed and demonstrated in the EuroView 2012 workshop. We showed that the SOA paradigm can be applied to networks by utilizing the concepts of self-contained building blocks, dynamic protocol graphs (PGs) and functional composition (FC) methods. We demonstrated that both short-term flexibility (i.e., networks are adapted based on application requirements) and long-term flexibility (i.e., networks can be evolved) can be achieved by using the architecture.

Keywords: SOA, Network Architecture, Selection & Composition, Template, AHP.

1 Introduction

The Internet today faces many challenges in terms of security, addressing, and mobility [9]. The problems originate from the architectural [23] issues of network functionality, their relationship, and the lack of design and evolution principles.

The Internet architecture is a layered system like TCP/IP stack based on OSI model. According to the OSI reference model, it should be possible to modify or even exchange the implementation of a layer without the need to adapt to the adjacent layers [19].

In today's practice, there are layer violations in the Internet, because of dependencies among protocols of different layers. For example, the addressing protocol in layer 3 (IPv6) requires an updated transport protocol (TCP) in layer 4. Thus, the evolution of the Internet "depends on rough consensus about technical proposals, and on running code" [2]. The Internet has become a complex system where it is hard to predict how the modification of one protocol affects the overall system. Many issues considered by the IETF IPv6 working group reflect this complexity [4].

As major changes in the Internet seem to be impossible, especially within a short time-frame, as a result new disruptive features are deployed in overlay networks. Overlays are usually designed only for a specific purpose such as filesharing, telephony, video-broadcasting, and are typically not open for arbitrary extensions or reuse. Hence, overlays are not a suitable alternative for a generic network infrastructure like the Internet.

Thus, there is a need of rethinking about network architectures [1]. This article presents the basic concepts of network architectures based on the service orientation paradigm from layered to layerless as shown in Fig. 1. The main goal is to develop a flexible network architecture which can be adapted to application requirements and network capabilities as well as to be able to integrate new functionalities easily.

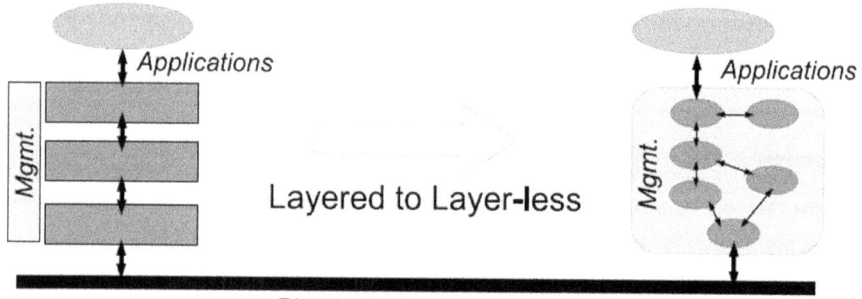

Figure 1 Layered to LayerLess Architecture

2 Service Oriented Network Architecture

Now the question is: how to apply the SOA paradigm to constitute a network architecture? The main element of SOA is a service. A service reflects the effects of an activity, i.e., a service represents a higher abstraction level since different algorithms may implement the same service. A building block is an implementation of a service. A Micro-protocol (MP) is an example of a building block such as a retransmission MP, a data encryption MP, and a Monitoring MP. Each building block can generate one or more effects, for instance, a retransmission BB has an effect of reliable data transmission, a data encryption BB has an effect of confidential data transmission. But there are also effects like increasing the end-to-end delay or reducing the maximum payload size. The interfaces of a building block should reflect the provided service and hide the implementation details. Building blocks should also use generic interfaces (i.e. as used in WSDL) so that interaction between building blocks does not require extra adapters.

A network architecture should be flexible in two ways. Firstly, networks should be able to adapt to specific customer or application needs and changing environmental conditions. Secondly, networks should be able to evolve, i.e. to add, change and even remove the functionality.

The flexibility is achieved by composing several (smaller) services to a more complex and customized service. In today's networks, complex protocols are organized in layers, building a static protocol stack, sometimes called Protocol Graph (PG) ([15]). Service oriented network architectures aim at supporting dynamic composition of services, i.e. dynamic PGs[1]. Without being dependent on a static PG it is easier to make use of new protocols (i.e., building blocks) and to reuse a functionality on different levels. Having dynamic PGs implies that there is no static placement of a functionality as defined by the layers of the OSI reference model. In this sense, such networks will be layerless so that compression or encryption can be used for application payload only or also for some protocol headers. Furthermore, it is not necessary that protocols are processed in a sequence, for example there might be different branches in a PG to handle different but related data types within a flow, e.g., signaling and media streaming. In order to enable dynamic PGs the interaction between the building blocks should not be defined by an executable code, but by the description which can be easily changed.

---
[1] corresponds to a workflow in SOA terminology

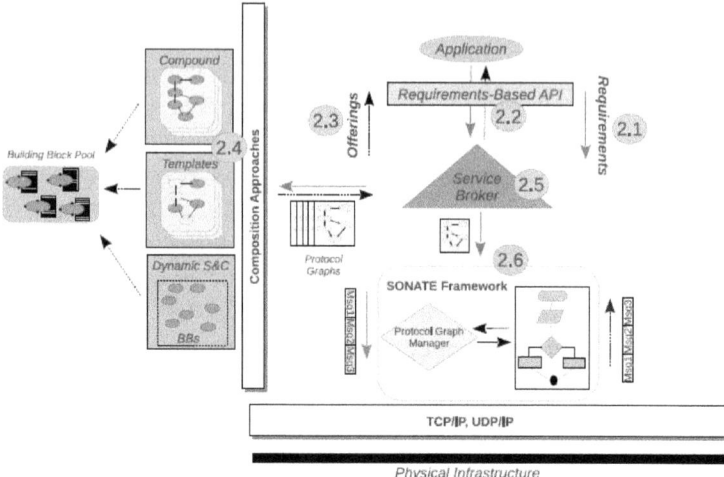

Figure 2 Proposed Network Architecture (the number in the blue circle indicates the subsection where the component is described in the paper)

Fig. 2 depicts the proposed approach. An application sends its requirements via the requirements-based API which is received by the service broker. Service broker forwards the requirements to the composition approaches where the PGs are generated. The composition process may generate more than one PGs so that the most suitable one can be selected by the selection process running inside the service broker. After selection of the most suitable PG, it will be directed to the SONATE framework for the execution and to initiate the communication between applications.

The following sub-sections describe the components of the architecture in a sequence: 2.1. application / user requirements, 2.2. requirements-based API, 2.3. network offerings, 2.4. a functional composition approach based on templates, 2.5. AHP-based service selection method in the Service Broker, and 2.6. the SONATE execution framework.

Subsection 2.7 briefly explains the prototype that has been demonstrated in the EuroView 2012 workshop. Standardization possibilities of the concept is discussed in Section 3. Section 4 concludes the paper.

2.1 Requirements

Requirements from a user or an application are represented by effects, operators, attributes, and weights as shown in Fig. 3 [13]. An effect is just the name, attributes quantify or qualify the effects and operators link effects

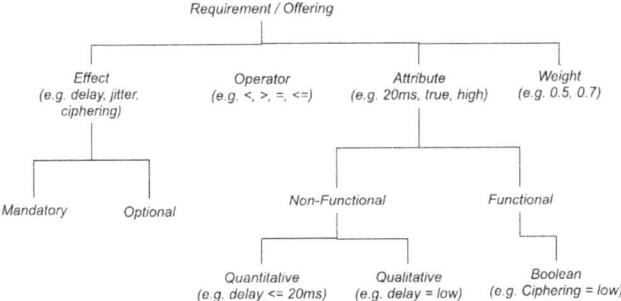

Figure 3 Requirements/Offerings

to attributes, typical operators are <, >, =, <=, >=, etc. Attributes can be represented in different ways by giving an exact quantity (e.g. *delay < 20ms*), Boolean values (e.g. *Packet ordering = true*) or in a qualitative way (e.g. *delay = low*), though qualitative parameters need extra mapping or definition, it may vary with respect to the context. The weight of an effect expresses the application's or user's priority over other effects.

2.2 Requirements-Based API

The existing API was made with the assumption that the Internet supports a limited number of protocols and relies on applications to specify the exact protocol to use. The current API hinders the deployment of new transport protocols such as SCTP [24], DCCP [6] and new addressing schemes as an application is obliged to specify an address type (IPv4 [17] or IPv6 [3]). Stipulation of protocol by applications fosters tightly bound coupling, which forces the network stack to use that exact protocol rather than an improved version of a protocol or one that is more suitable with respect to the network conditions. In order to deploy a new protocol in the Internet architecture, it is not enough just to change the application but it also requires modifying the API or using another API so that a user can unveil its intention to use the newly deployed protocol.

An application needs to be modified if it wants to use TCP [18] or UDP [16]. Peer addressing and address-resolution are part of an application, which makes an application address type dependent. Different addressing types require different socket so that there is a difference between the IPv4 TCP socket structure and the IPv6 TCP socket structure and same is true for UDP.

The call of "setsockopt" is an example of another dependency where protocol specific options such as TCP_NODELAY can be turned on or off, as options are specific to a protocol so that it is a must for an application to know details about a protocol.

Currently, there are multiple existing APIs each developed for different transport protocols. If an application needs to switch a transport protocol, it is not enough just to adjust socket options or to change addressing family but it is also required to use a particular API given for a particular transport protocol.

Abstraction is used for hiding the complexity and to encourage the flexibility. In our approach, we propose an API by which an application can send its requirements in an abstract form to the underlying network such that applications do not need to rely on specific protocols; the process of selection would rather be handled by the network architecture. This triggers the requirement for the network architecture to be able to handle those abstract requirements from applications. Abstract requirements from applications also help to create an unified API so that a single API can be used for multiple transport protocols.

Current applications are tightly coupled with the given protocols, though they only care about whether its connectivity demands are fulfilled. A requirements based API [22] will alleviate the developer from choosing a protocol or even a protocol specification. Instead, requirements will be communicated to the underlying network architecture. Using these requirements, the explicit connection characteristics are requested for the new communication relationship. Requirements are specified in terms of effects / capabilities, an effect is a visible outcome of a functionality such as flow control functionality provides effect of transmission rate adaptation between two parties.

2.3 Network Offerings

To constitute a PG based on the application requirements, the offered services from the building blocks needs to be described so that the most suitable BB can be selected and composed. Moreover, the services of the PGs should also be described so that the best PG based on application requirements can be selected and used for communication. For describing these services, a communication service description language was developed. The language consists of a taxonomy of vocabularies and a grammar [13]. The details of the description language is beyond the scope of this paper.

2.4 Functional Composition

Functional Composition (FC) is the process of selecting and binding of the building blocks (BBs). The following sub-section describes the template-based functional composition approach which is developed and demonstrated in the EuroView 2012 workshop [8] under the umbrella of a German-based Future Internet (FI) research and experimentation project named G-Lab [7].

2.4.1 Template-based functional composition

In order to create a requirements-based PG out of given functionalities, it is necessary to define data and control flow of selected functionalities. The control flow in a PG is defined by the placement of functionalities while the data flow is defined by the connections among functionalities. The template-based composition is a partial-runtime approach where ordering of functionalities and their connections are defined at the design-time.

The basic idea of the approach is to split the functional composition process among different time-phases (i.e. design-time, deployment-time, and run-time) so that relatively inefficient activity in terms of time is performed at design time and potentially less time consuming activities are performed at run-time. In this case, time consuming activities are the selection and the placement of functionalities but not the actual building blocks (i.e. selection of encryption and compression functionalities but not their implementations/BBs) and placing them in an appropriate order (i.e., encryption is placed on the top of compression) in addition to connect them so that they can interact with each other. To utilize the less time critical epoch (e.g. design-time) and yet to provide flexibility, the template based composition approach utilizes the devised abstraction of place-holder instead of using actual functionality as shown in Fig. 4.

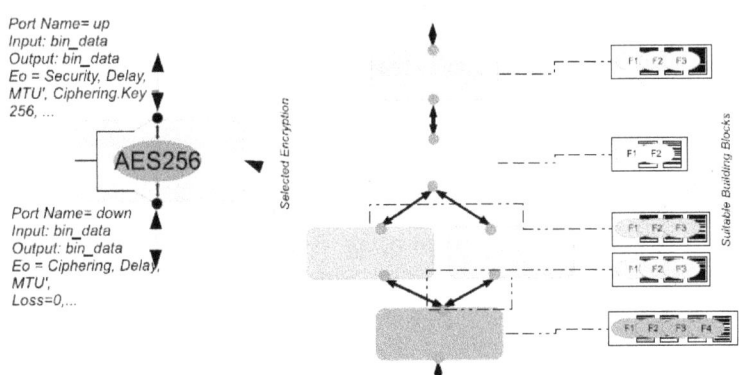

Figure 4 Template and Placeholder

The Place-holder is one of the major entities in this approach. The Place-holder provides named endpoints so called ports. The ports are well defined in terms of effects or capabilities that must be provided by a place-holder. Effects can be differentiated by the offered (i.e. provided by the port) and the required (i.e. accepted by the port) effects as shown in Fig. 4. An example effect of an encryption functionality is ciphering and this effect is covered by various mechanisms such as AES, DES, Blowfish, etc. so that it should be possible to change the mechanism without changing the effect. Aforementioned mechanisms can have many implementations and may differ in terms of defined ports (i.e. it implies same covered effects on those ports). Any implementation which also has same ports as described in a place-holder can be a suitable match to fill that place-holder.

code.1 Placeholders Example

```
<Placeholders>
    <PlaceHolder Name="Encryption" ID="1">
    <ToggleEnable isEnable="true"/>
    <Port PortID="up">
        <OfferedEffect Effect="Ciphering.key" Operator="=" ↵
           Attribute="256"/>
        <OfferedEffect Effect="Delay" Operator="=" Attributed ↵
           ="1ms"/>
    </Port>
    <Port PortID="down">
        <OfferedEffect Effect="Ciphering" Operator="=" ↵
           Attribute="true"X/Of f eredEffect >
        <OfferedEffect Effect="Loss" Operator="=" Attributes↵
           ="0"> </OfferedEffect >
    </Port>
    </PlaceHolder>
</Placeholders>
```

The Place-holder also contains ports for general functionalities such as management, administration and monitoring. But those ports are active only when a selected BB also provides that data, hence these ports are optional and not considered in the BB selection process.

Template Description Language: The template description is split in four main parts so called Domains, Placeholders, Connections and CoveredEffects. Where, Domains section describes the types of domains that are covered by a template, examples of domains are telephony, video streaming, file transmission, etc. The Placeholders section describes the covered functionalities in a template, which is further sub-divided into individual placeholder and its ports. Connections section of the language deals with the ordering and the connections of the place-holders.

Example of an encryption placeholder is shown in the code.1, which provides two offered effects (i.e. Ciphering.key and Delay) through the port "up" and two offered effects (i.e. Ciphering and Loss) through the port "down". In this example no required effects are described as this functionality accepts binary data as an input and gives binary data as an output.

Template Selection: For the selection of a template two simple matching algorithms have been implemented. After receiving the application requirements, the domain will be extracted and it will be checked against stored domain policies at the system. The domain policies are required in order to have the special constraints/requirements which are not provided by an application such as compression in file transfer (i.e. an application does not care if data will be compressed or not as long as a file is efficiently transferred to the communicating partner). After retrieving the domain policies, it will be merged with the application requirements to find a matching template.

The first matching process goes through all of the merged requirements and check against covered effects in a template. This selection process works separately for a single template. The advantage of this approach is that multiple processes can run in parallel without being dependent on each other.

In the second algorithm, the selection process reads first the requirement and checks against the available templates at the system and separates the one which fulfills the requirement. After that it reads the second requirement out of the requirements list and examines against sorted out templates from the previous step. And this process goes on until one or more than one templates have been sorted out which fulfill the given requirements. This method is useful for the devices with limited cache as memory switching in threading can be costly. When more than one templates satisfy the requirements then multi-criteria decision analysis algorithms can be used to select the most optimal template.

Selection of BBs for Filling a Template: There can be more than one BBs which fulfill the application requirements and network and other constraints so called "suitable" building blocks. If an optimal choice has to be made then all possible workflows are generated. Based on the qualitative parameters, the best PG is selected. The Analytic Hierarchy Process (AHP) is used to select the best PG which is further described in the following section.

2.5 Service Broker

The Service Broker is responsible for selecting protocol graphs (PGs) with respect to the application requirements. Selecting the best PG using a single selection criterion is trivial. For example, if there are two PGs where one offers 100ms end-to-end delay and another offers 200ms, then we should obviously select the one with less delay.

However, the communication services provided by PGs have multiple selection criteria such as delay, throughput, loss ratio, jitter and cost. That is why, selecting the best PG is a Multi-Criteria Decision Making problem (MCDM). For solving such a problem, several Multi-Criteria Decision Analysis (MCDA) approaches are used in managerial science like Analytic Hierarchy Process (AHP), ELECTREIII, Evamix, Multiple Attribute Utility Theory (MAUT), Multi - Objective - Programming (MOP), Goal Programming (GP), NAIADE and Regime [5].

We used AHP to select the best service for two reasons, firstly, it uses an absolute scale to derive priorities that also belong to the relative absolute scale (like probabilities) that can be combined like the real number system. Secondly, there is a way to check the consistency of the evaluation measures.

2.5.1 Adaptation of Analytic Hierarchy Process (AHP) for service selection

The Analytic Hierarchy Process (AHP) needs to be adapted for selecting the best communication service automatically.

AHP is a process designed for assisting human decision making which is used in many application areas like social, personal, education, manufacturing, political, engineering, industry and government [20]. Basically, AHP is used for determining priorities of different alternatives. The details of the AHP process is beyond the scope of this text.

To use AHP in PG selection, the following steps are performed

1. Define the goal and the selection criteria for achieving the goal
2. Priority assignment of the selection criteria as an application requirement
3. Priority assignment of the criteria for the offered services

The first step is to define the goal, which is to select the best communication service, and the selection criteria to achieve that goal. The selection criteria are actually a set of required effects. Examples of the selection criteria are delay, throughput, loss rate, jitter, MTU and cost. Both functional and non-functional criteria can be selected.

After determining the selection criteria, the next step is to assign pairwise priority between the selection criteria. One of the reasons of pairwise priority assignment is that, it is easier for a person to take two criteria and to assign a priority one over the other. It is initially difficult for a new application developer to assign a pairwise priority. But, the efficiency of the priority assignment process can be improved with the experience of the application developer.

The third step of the process is to assign pairwise priority between the offered services based on those selection criteria. However, as pairwise priority assignment is a time-consuming task, and as offerings are decoupled from the applications, the pairwise priority assignment of the offered services based on those selection criteria needs to automated.

This requires a mapping mechanism to map the measured/calculated values of the offered services to the pairwise priority assignment scale which will be discussed in the next section.

The priority vector coming from the application side is then multiplied by the priority vector from the offering side. The result is then called the overall priority vector. The service with the highest priority value in the overall priority vector is the best service.

2.5.2 Automated priority assignment for the offerings

Different PGs can have different effects. The value (or attribute) of these effects can be assigned beforehand based on benchmarks or can be obtained dynamically by using a sensing software. Whichever way the attributes are obtained, the offered effects need to be automatically prioritized as the offerings are decoupled/hidden from the application. Therefore, an automatic The mapping should have certain properties. First, the mapping must be generic, i.e. not specific to effects or units of measured values. Second, the mapping must be monotonic.

An approach for mapping has been proposed which uses a monotonic interpolation/extrapolation scheme [12] as shown in Fig. 5. In this case, the application requirements provide value points for interpolation/extrapolation (must be monotonic) of measured values to the priority scale. A monotonic interpolation/extrapolation of these points is used to define a mapping. In addition, the specific measured values of the offerings are then mapped to these priorities. Assuming that $f()$ is a function used to define a mapping. As

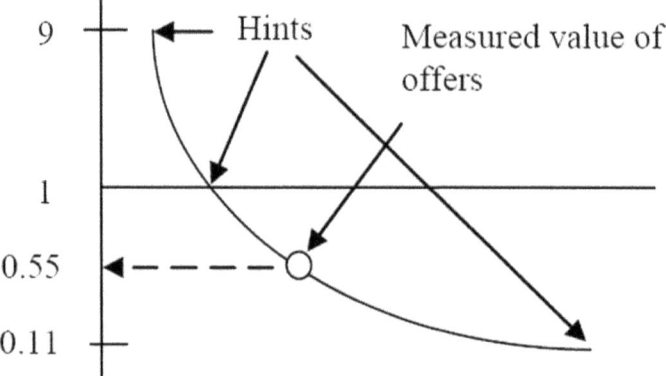

Figure 5 Mapping mechanism

an example, considering interpolation, the requirements must contain at least the following two points

- x_0, where $f(x_0) = 0.11$
- x_n, where $f(x_n) = 9$

If there are measurement values, y, not within the interval $[x_0, x_n]$, we can extrapolate

- if $y < x_0$, then $f(y) = 0.11$
- if $y > x_n$, then $f(y) = 9$

To use inter-/extrapolation, an application developer must specify two points but can have as many parameters as he wants to be more precise.

The aforementioned mapping mechanism is used to assign a priority of one service over another for every selection criteria (effect).

2.6 SONATE Execution Framework

The framework for service oriented network architecture (SONATE) executes the selected PG. The flow consists of operations to invoke the specific BB and to pass the corresponding values as input parameters to the connected BBs.

The PG is described in an Extensible Markup Language (XML). The SONATE framework has the ability to process the XML based PG. The processing of PG involves retrieval of BBs from the repository, connection of BBs as specified in the PG and execution of BBs in the given order.

The SONATE framework has been implemented using the JAVA programming language. The "Port concept" [21] has been used for the interaction among building blocks. A crucial point in maintaining flexibility is loose coupling. The building blocks that are tied to specific implementation details of other building blocks, reduce the flexibility to freely combine building blocks. To enforce loose coupling, the BB interaction model hides all BB's implementation details from each other.

In the runtime composition approach, BB instances need to communicate with each others to make a PG, the concept of named communication endpoints called "ports" is introduced. Building blocks can use the ports to distinguish different kinds of communication partners or different operation modes.

2.7 Prototype Demonstrated

We demonstrated the concept of service oriented network architectures in the EuroView 2012 workshop [8]. More specifically, we showed how networks react to user's requirements, network/administrator constraints by dynamically selecting and composing a customized protocol graph.

In the demonstration, the Firefox browser retrieves different types of data (Voice, Image) with different user requirements, network constraints and domain policies.

Two servers with different network conditions such as bandwidth and jitter were configured. We have extended a Firefox plugin, to interconnect Firefox with the Requirement-based API to communicate with the components in the SONATE Framework. Firefox sends the requirements via API to the service broker, which is responsible for selecting the best PG at the end. The service broker forwards requirements further to the template based composition engine, which creates and returns multiple PGs so that the most suitable is selected by the service broker using AHP. We encoded all application requirements, network constraints and administrative policies in the URL so that they can be uniquely identified. The selected PGs are displayed by a browser-based visualizer.

We showed that both long and short term flexibility are achieved by the proposed architecture.

Short Term Flexibility: In this scenario, different image compression mechanisms have been provided which are selected based on the application requirements and the network conditions. Application requirements provide trade-off between expected quality and transmission speed. If an application requires a better quality then no image compression or a very low compression mechanism is deployed. But, if transmission speed is the priority of a user then the best compression mechanism (i.e., high compression ratio) is selected. The scenario is complicated by the given administrative policies with respect to network conditions. If the given bandwidth is less than or equal to certain threshold like 1 Mbps then compression mechanism is deployed, otherwise, in the case of faster networks, no compression mechanism is deployed in the PG.

Long Term Flexibility: In this scenario, a better implementation of an encryption mechanism has been deployed in the system which provides better security in terms of key-strength. As soon as a user demands for higher security, the newly deployed mechanism is used in the automatically generated PG. The scenario shows how quickly a newly introduced functionality can be deployed in the presented architecture.

3 Standardization Candidates

The components of the architecture including communication service description language, requirement-based API, and template-based functional composition can be seen as the potential candidates for standardization. These items can be included in the ITU-T study group 13 (ITU-T SG 13) where the potential future networks technologies like cloud computing, mobile, and next-generation networks are discussed [11]. Moreover, the API can be examined within the Name-Based Sockets Architecture community of IETF [10]. In addition, the language can be considered in the Operations and Management Area of IETF [14].

4 Conclusion and Future Work

The implementation of communication protocols in the current Internet architecture is not loosely-coupled which hinders the evolution of the Internet. This article describes how the principles of Service Oriented Architecture (SOA) can be employed to develop a flexible network architecture. The SOA paradigm can be applied to networks by utilizing the concepts of self-contained building blocks, dynamic protocol graphs (PGs) and functional composition (FC) methods. It is shown that both short-term flexibility (i.e., networks are adapted based on application requirements) and long-term flexibility (i.e., networks can be evolved) can be achieved by using the architecture.

Heterogeneity, availability of diverse functionalities in different network elements, is an issue which needs to be tackled in the future. The proposed approach expects to have a controlled environment where every node has the same functionalities available. However, existing mechanisms like negotiation can be used to deal with heterogeneity. Moreover, new functionalities can be deployed from a trusted domain.

References

1. R. Braden, D. D. Clark, S. Shenker, and J. Wroclawski. Developing a next-generation internet architecture. 2000.
2. B. Carpenter. Architectural Principles of the Internet. RFC 1958 (Informational), June 1996. Updated by RFC 3439.
3. S. Deering and R. Hinden. Internet protocol, version 6 (ipv6). *RFC2460*, Dec 1998.
4. Ralph Droms and Jari Arkko. Ipv6 status pages, March 2008.
5. Matthias Ehrgott and Xavier Gandibleux. Multiple criteria optimization state of the art annotated bibliographic surveys. *Kluwer Academic Publishers*, 2003.
6. S. Floyd and E. Kohler. Profile for datagram congestion control protocol (dccp) congestion control id 2: Tcp-like congestion control. *RFC4341*, March 2006.
7. German lab (g-lab). http://www.german-lab.de/. Online; accessed 11-October-2012.
8. Daniel Günther, Dennis Schwerdel, Abbas Siddiqui, Rahamatullah Khondoker, Bernd Reuther, and Paul Müller. Selecting and composing requirement aware protocol graphs with sonate. *In 12th Würzburg Workshop on IP: Joint ITG, ITC, and EuroNF Workshop on 'Visions of Future Generation Networks' EuroView 2012*, July 2012.
9. Mark Handley. Why the internet only just works. *BT Technology Journal*, 24(3), 2006.
10. Name-based sockets architecture draft-ubillos-name-based-sockets-03. http://tools.ietf.org/html/draft-ubillos-name-based-sockets-03. Online; accessed 05-May-2013.
11. Itu-t sg13: Future networks including cloud computing, mobile and next-generation networks. http://www.itu.int/en/ITU-T/studygroups/2013–2016/13/Pages/ default.aspx. Online; accessed 05-May-2013.
12. Rahamatullah Khondoker, Bernd Reuther, Dennis Schwerdel, Abbas Siddiqui, and Paul Müller. Describing and selecting communication services in a service oriented network architecture. *In the proceedings of the 2010 ITU-T Kleidoscope event, Beyond the Internet? Innovations for future networks and services, Pune, India*, December 2010.

13. Rahamatullah Khondoker, Eric MSP Veith, and Paul Müller. A description language for communication services of future network architectures. *In Proceedings of the 2011 International Conference on the Network of the Future*, pages 69–76, 2011.
14. Operations and management area. https://datatracker.ietf.org/wg/.Online; accessed 05-May-2013.
15. Sean W. O'Malley and Larry L. Peterson. A dynamic network architecture. *ACM Transactions on Computer Systems*, 10:110–143, 1992.
16. J. Postel. User datagram protocol. *RFC768*, Aug 1980.
17. Jon Postel. Darpa internet program. *RFC791*, Sept 1981.
18. Jon Postel. Transmission control protocol. *RFC793*, Sept 1981.
19. Recommendation. Recommendation x.200 (07/94), x.200 information technology, open systems interconnection, basic reference model the basic model. 1994.
20. Thomas L. Saaty. Decision making with the analytic hierarchy process. *Int. J. Services Sciences*, 1(1):83–98, 2008.
21. Dennis Schwerdel, Danile Günther, and Rahamatullah Khondoker. A building block interaction model for flexible future internet architectures. *7th EURO-NF CONFERENCE ON NEXT GENERATION INTERNET, 2011*.
22. Abbas Siddiqui and Paul Müller. A requirement-based socket api for a transition to future internet architectures. *Sixth International Conference on Innovative Mobile and Internet Services in Ubiquitous Computing (IMIS-2012)*, 2012.
23. IEEE Std. Ieee std., 1471-2000 ieee recommended practice for architectural description of software-intensive systems-description. 2000.
24. R. Stewart, Q. Xie, and et al. Stream control transmission protocol. *RFC2960*, Oct 2000.

Biographies

Abbas Siddiqui is a Ph.D. candidate with the topic of Flexibility in the Network Architectures, where he proposed a partial-runtime composition approach to create customized network-stacks. The name of his approach is "Template-Based Composition". After completion of his master in "Electrical & Communication Engineering" with focus on Mobile & Internet Engineering from University of Kassel, Germany, he started to work as a software engineer in the industry, where he gathered several years of experience as a software developer before, leaving the industry for the sake of research. His current research revolves around Service Architectures, Smart Living, e-Health, and Sensors Technology.

Paul Müller is a Computer Science (CS) professor and director of the regional computing center at the University of Kaiserslautern in Germany. His current research interests are mainly focused on distributed systems, Future Internet (FI), and service-oriented architectures. His research group Integrated Communications Systems(ICSY) is aiming at the development of services to implement integrated communication within heterogeneous environments especially in the context of the emerging discussion about FI. This is achieved by using concepts from service-oriented architectures (SOA), Grid technology, and communication middleware within a variety of application scenarios ranging from personal communication (multimedia) to ubiquitous computing.

Dr. Kpatcha Bayarou received his Diploma in electrical engineering/automation engineering in 1989, a Diploma in computer science in 1997, and his Doctoral degree in computer science in 2001, all from the University of Bremen in Germany. He joined the Fraunhofer Institute for Secure Information Technology (Fraunhofer SIT) in 2001. He is the head of the "Mobile Networks" department that focuses on Cyber Physical Systems and Future Internet including vehicular communication. Dr. Bayarou managed several EU and nationally funded projects and published several conference papers related to security engineering of mobile communication systems, mobile network technology, and NGN (Next Generation Networks).

Since 2010, Rahamatullah Khondoker has been working towards his PhD degree on "Description and Selection of Communication Services for Service Oriented Network Architectures (SONATE)" at the University of Kaiserslautern in Germany. He was awarded from Ericsson, Germany in the year 2008 and from the FIA Research Roadmap group in October 2011. Currently, he is affiliated with the Fraunhofer SIT located in Darmstadt, Germany. He worked with the DFG project (PoSSuM), BMBF projects (G-Lab, G-Lab DEEP, Future-IN), and EU projects (PROMISE, EuroNF). Currently, he is focusing on the security of Future Internet Architectures, Software-Defined Networking (SDN), and Network Function Virtualization (NFV).

Towards a Framework for Health IT Standardization in Mexico

Arturo Serrano-Santoyo[1], Veronica Rojas-Mendizabal[1], Roberto Conte-Galvan[1], Amanda Gomez-Gonzalez[2] and Angelica Baptista Silva[3]

[1]CICESE Research Center, Ensenada, Baja California, Mexico,
serrano@cicese.edu.mx, vrojas@cicese.edu.mx, conte@cicese.m
[2]Mexican Space Agency, gomez.aem@gmail.com
[3]Oswaldo Cruz Foundation, Brasil, silva.angelica@gmail.com

Received: 29 March, 2013; Accepted: 8 October, 2013

Abstract

The increasing penetration of ICT in the health sector, in conjunction with the explosion of converging applications covering biotechnology, nanotechnology and green technologies, is creating important opportunities for developing efficient systems and processes that can improve the quality and coverage of health services in developing countries. The pace of technology growth and the lack of comprehensive technology adoption programs and a sound regulatory environment present important challenges if these countries are to take full advantage of the new international scenario of digitalization and global convergence. The role of standardization in a complex and highly converging context is particularly relevant. In consequence, national ICT regulatory administrations in emerging economies are committed to consolidating and strengthening their standardization processes and policies, and national health institutions acknowledge the need for standards that include information technology elements in medical practices. This article delineates the basic elements for the development of a comprehensive framework for Health IT standardization in Mexico. We take into account the multidimensional and complex nature of the standardization process and the importance of interactions among all the actors involved. Our proposal stems from the establishment of a National

Program for eHealth in Mexico as recommended by both the WHO and the ITU.

Keywords: Health IT, standardization, innovation, ICT regulation.

1 Introduction

Information and communications technologies (ICT) play a pivotal role in the conformation of health ecosystems worldwide. The delivery of health services, the implementation of data security and privacy programs and the development of medical devices and future services greatly depend on the deployment of efficient cyber infrastructures. Since the last decade, joint and independent efforts on the part of the World Health Organization (WHO) and the International Telecommunications Union (ITU) have resulted in a series of recommendations to all countries regarding the integration of ICT in their national health information systems and health infrastructure [1, 2]. The aim is to support member states with the vision and resources for the development or revitalization of their Health IT policies and strategies, focusing on key emerging technologies such as mobile health (mHealth), telemedicine, eLearning, management of patient information, legal frameworks, safety and security on the internet. Both the WHO and the ITU consider Health IT standardization and interoperability issues as fundamental components for accomplishing a sound national strategy for Health IT, urging member states to structure roadmaps for the implementation of health data standards and to ensure compliance with and adoption of such standards in the public and private sectors [3].

Recognizing the existing disparity between developed and developing countries in relation to standards development, the ITU has focused special attention on reducing this divide. To this aim, the program "Bridging the Standardization Gap" was established in 2009 [4], one objective of which is to identify and understand the gaps that inhibit the process of standards development and implementation in developing countries. At the same time, the ITU acknowledges that the standards capability of a country represents a significant element affecting the digital divide between developed and developing nations.

In light of the above-mentioned concerns, reviewing and upgrading public policies of ICT standardization in developing countries is essential for accelerating their transit toward the knowledge economy, while a diagnosis of the national standards capacity is an important element for assessing and identifying each country's primary standardization gaps. As part of the "Bridging the Standardization Gap" program, the ITU developed the Tool for Assessing Standards Capability (TASC) to evaluate national standards readiness in the developing world [5]. The application of this tool provides key indicators for the development of recommendations and best practices for capacity building as well as for the establishment of public policies geared to reducing the existing standardization gap.

Health IT standardization calls for diverse and tailored solutions for each country, involving complex and multidimensional issues in which the understanding and interplay of health, trade and intellectual property are necessary for sound policy making. Likewise, a balance between regulation and innovation is conducive to the configuration of public policies that capitalize on the opportunities for cross-fertilization among the areas involved in Health IT standardization. In this scenario, it is important to acknowledge that innovation in health technologies differs from innovation in general due to the ethical dimension of medical research, the national health regulatory framework and the liabilities, costs and risks involved [6]. Consideration of these issues entails an interdisciplinary approach in the development of a comprehensive Health IT standardization framework.

Throughout the article we use a conceptualization of standardization as the process of creating, implementing or using a standard (a concept or realization based on common agreements or rules based on set theory) [7]. From this perspective, we consider the case of Health IT standardization, covering the use of ICT in health information systems, infrastructures, equipment, devices, processes, and the delivery of health services. This paper is organized as follows: Section 2 presents a summary of the Mexican Health IT context, highlighting some key indicators, issues and challenges; Section 3 describes the elements of a framework for Health IT standardization in Mexico with reference to the WHO and ITU guidelines and the current policies of health agencies in the Mexican health ecosystem; in Section 4, we identify major challenges that, in our view, Mexico—as well as other emerging economies—faces to develop a sound and sustainable Health IT standardization framework. Our final remarks and conclusions are presented in Section 5.

2 The Mexican Context of Health IT Standardization

With a population of around 114.9 million and a GDP of 1.153 trillion (US dollars), Mexico is the second largest economy in Latin America and member of the G20 since 2008 [8]. Given its size in the global economy, the strengthening and consolidation of Mexico's regulatory and standardization frameworks could give rise to significant opportunities for improvements in productivity, competitiveness and innovation in key advanced manufacturing and service sectors. Considering the transversal nature of ICT, reducing the current standardization gap emerges as a significant step towards positioning the country as an important player in the global economy.

With respect to the Mexican health sector, government agencies have made significant efforts to implement strategies and actions that would increase the role of ICT in practically all the links of the value chain of the national Health IT ecosystem [9, 10]. Furthermore, the ICT regulatory bodies have acknowledged the importance of standardization for advancing the competitiveness indicators of the country; at the same time, the national health institutions recognize the need for standards that include ICT in their medical practices, delivery of services and equipment.

Current governmental and legal structures create a foundation for the development of a comprehensive Health IT standardization framework. The following initiatives are noteworthy:

- Establishment of a national platform for standardization in the health sector, comprising agencies, instruments for medical attention, processes and capacity building [11]. The General Directorate of Quality and Health Education, an agency of the Mexican Ministry of Health (SSA-acronym in Spanish), is responsible for coordinating efforts on health standardization processes in the country (see Figure 1).
- Creation of the National Center for Health Technology Excellence (CENETEC-acronym in Spanish). This Center was created under the auspices of the SSA to produce and disseminate information and the use of medical technologies, based on evidence of their safety, effectiveness and efficiency, for the benefit of the population and the advancement of medical practice [12].
- Publication of two key regulatory instruments: The Mexican Electronic Health Records Standard and the country's Federal Law for Personal Data Protection [13, 14].
- Establishment and strengthening of the E-Mexico National System. This national program was created in December 2000 to propel the country toward the information society [15]. The program is coordinated by the Mexican Ministry of Communications and Transport (SCT-acronym in Spanish).

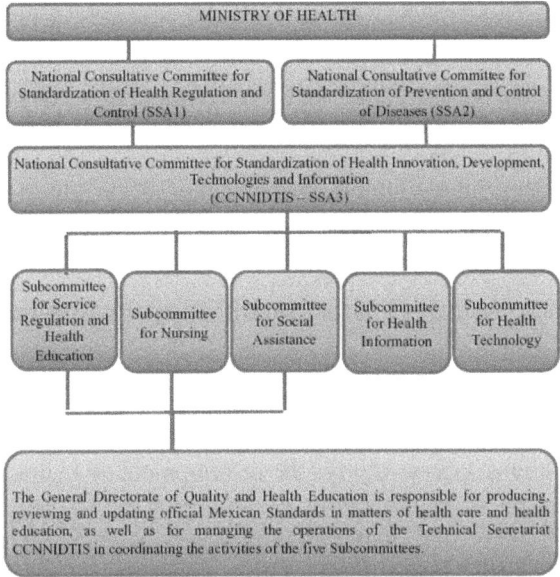

Figure 1 Mexican Health Standardization Structure (adopted from [11]).

Despite the efforts of government and private entities to support, increase and systematize the use of ICT in the health sector, Health IT standardization matters require more sustained attention and strategic actions. The following is a general overview of issues related to the current status of Health IT standardization:

- Results of the application of the ITU-TASC tool for assessing standardization capabilities indicate that the country faces major challenges in standardization matters [5].
- The e-Health country profile for Mexico of the WHO Global Observatory for e-Health shows a significant deficit in key indicators, particularly those related to Health IT standardization [16].

Mexico has a considerable foundation from which it can advance to a more comprehensive and focused participation in Health IT standards development. The strategies and current regulatory structure of the health sector, however, reflect an orientation geared towards simply meeting technology users' needs for standardization. That is, there is a lack of focus on the innovation and technology development that could boost local participation in the creation of standards related to systems, processes and next generation medical devices (including mHealth). Although there is a substantial platform of standards covering different areas of medical practice, Health IT standards development requires concentrated strategies and actions in order to reduce the gap and improve the status of Health IT standardization in the country. More important, however, is that key actors in the Mexican health, ICT and regulatory environments acquire an understanding and awareness of the complex and interdisciplinary nature of standardization and its crucial role in the global scenario.

3 Elements of a Health IT Standardization Framework for Mexico

The pressing need for improving the quality and coverage of medical services in developing countries, particularly in remote and under-served locations, represents an important opportunity for the development of a comprehensive framework for Health IT standardization. We argue that this framework should adopt an interdisciplinary and collaborative approach and that Mexican health and ICT regulatory bodies need to take into account the importance of the development of endemic Health IT standards rather than merely focusing on their use and implementation. Moreover, we also posit that the process of standards creation emerges from the interactions of the four subsystems forming the Health IT ecosystem. This ecosystem can be viewed as an open socio-technical-legal system. Figure 2 shows the four subsystems considered

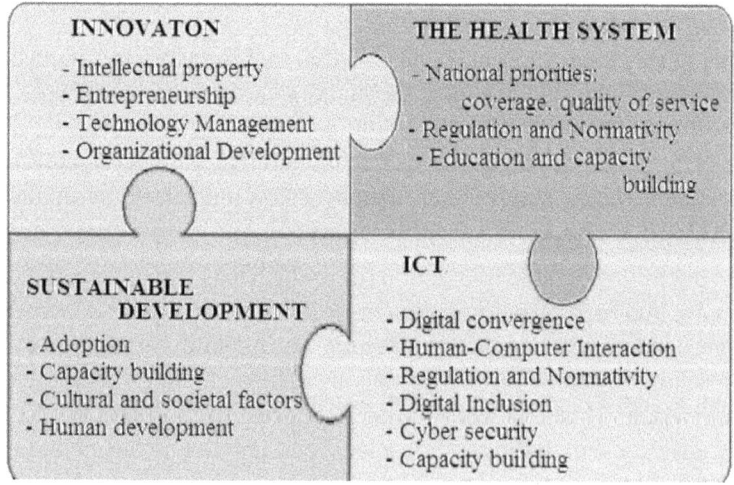

Figure 2 The Health IT Ecosystem.

in our approach: the innovation subsystem, the national health subsystem, the sustainable development subsystem and the ICT subsystem.

We also believe that a sound Health IT standardization process requires strategic and integrated action at the national level, making use of existing capacity while providing a solid foundation for investment and innovation, as suggested in the National eHealth Strategy Toolkit [2] developed jointly by the WHO and the ITU. We maintain that the application of this resource requires the interaction of actors with experience in strategic planning, communication and collaborative work.

Figure 3 shows the components of the National eHealth Strategy Toolkit and identifies the critical steps in each one of its three stages. The use of this Toolkit and the end result depend on the unity of vision of all stakeholders in regard to the critical issues involved in Health IT standardization and the successful orchestration of interactions among the subsystems described in Figure 2. The World Health Organization uses the term eHealth, which we consider to be the equivalent of the term employed by the ITU, Health IT. In this paper we have adopted the ITU term, Health IT, and only use the term eHealth for referencing purposes.

The National Health IT strategy for Mexico functions as the foundation for the development of our proposal. In our view, a program for Health IT standardization should arise from a sound national strategic plan with the participation of governmental, industrial and academic stakeholders. We

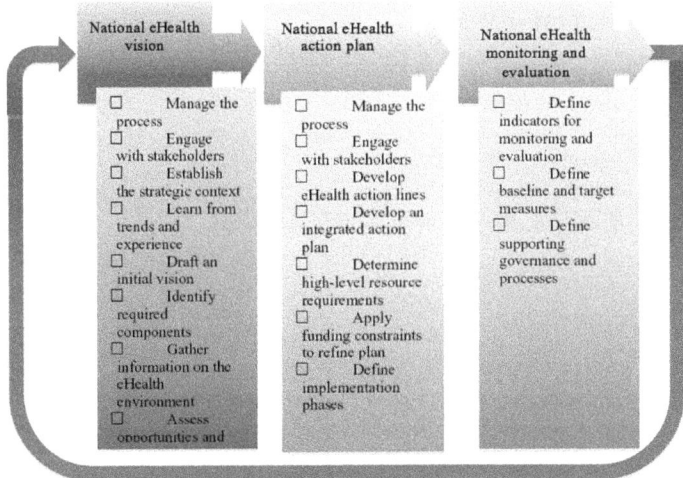

Figure 3 Toolkit for developing a National eHealth Strategy (adopted from [2]).

propose concrete steps for the development of a framework for Health IT standardization; Figure 4 describes the elements of our approach. We build our proposal around the creation and use of flexible standards [17] that can adapt to the frequent changes that are characteristic of the complex health environment. The flexible standards approach has demonstrated its usefulness when applied in various developing countries. Viewed from this perspective, standards can be created and maintained as complex adaptive systems where flexibility in use and scalability are essential for achieving the purpose of the standard. The authors of this flexible standards strategy argue that standards must be simple and easy to change and at the same time support a wide range of work practices, thus enabling radical change through small steps. Based on the flexible standards strategy, we propose that the process of Health IT standardization in Mexico initiate with the selection of one particular priority standard. The experience gained in the implementation of this standard would allow us to scale higher levels of complexity, including other standards and other requirements of the Health IT ecosystem as suggested by the above-mentioned authors [17].

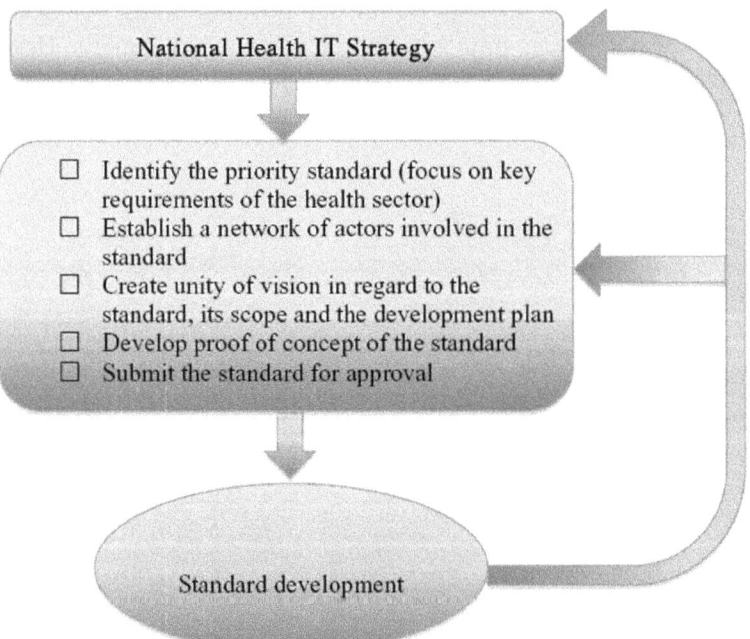

Figure 4 Elements of a framework for Health IT standardization.

4 Identifying Major Challenges in the Development of a Health IT Standardization Framework

As was mentioned in Section 2, CENETEC is a key institution in the Mexican Health IT ecosystem. CENETEC oversees the standardization of medical equipment and systems involving the use of technology. It is therefore the de facto organism in charge of Health IT standardization in the country. Since its inception in 2004, CENETEC has been the liaison with international standards bodies and has participated in the development of important documents for the WHO and other authorities working in Health IT and related fields [18, 19]. CENETEC's contributions to standardization matters in Mexico can be summarized as follows:

- Development of Mexican normativity (NOM, Mexican Official Standards) for:
    a) Generic medical instruments
    b) Medical devices
    c) Infrastructure and equipment requirements for medical facilities
- Development of guidelines for technical specifications.
- Development of guidelines and models for medical equipment.
- Development of guidelines for the implementation of Telehealth projects (infrastructure and equipment)
- Development of guidelines for clinical practices

According to CENETEC [20], it is necessary to address priority issues in the conformation of a sound and comprehensive standardization framework for Health IT. Some of the issues and recommendations are:

- Intensify the scope and coverage of standardization in all healthcare services of public and private medical institutions.
- Identify the cost-benefit, cost-impact and cost-opportunity indicators for Health IT.
- There is a widespread lack of "digital skills" and "informatics and telecommunications culture" at all levels of the Health IT ecosystem. This condition applies to medical practitioners as well as to decision-makers and management personnel.
- Some medical practitioners consider emerging technologies as a threat.
- There is a lack of a legal and ethics framework for Health IT.

The authors acknowledge the important work and contributions of CENETEC to the Mexican Health IT context. We suggest that in order to capitalize on the benefits of Health IT as well as achieve a comprehensive participation of Mexican institutions in standards creation, some remaining barriers should be eliminated:

- Infrastructure in rural and in some suburban locations is largely insufficient to utilize the most current Internet technologies.
- Even when basic infrastructure is in place, widespread interoperability standards for software are lacking.
- Local skills, knowledge, and resources may also limit the application of Health IT devices, equipment and systems.
- Failing to account for incompatible cultural subsystems that prevent the transfer of knowledge from one cultural context to another is a major contributing factor to Health IT failure. Without a good understanding of the local context, it may be difficult to integrate Health IT in a useful way.

Additionally, we posit that in order to develop a national program for Health IT standards creation, the consideration of the following factors is essential:

1. Development of a detailed analysis of the Mexican Health IT ecosystem including the interactions of the major agents in the strategic, tactic and operational levels.
2. Alignment of the national program for Health IT standardization in Mexico with the WHO strategic plan for 2012–2017 [21] as described in Figure 3.2.

With respect to the Mexican Health IT ecosystem, we acknowledge additional challenges such as:

1. The aspect of culture: Emerging economies such as Mexico face challenges in their efforts to increase competitiveness and productivity indicators. As mentioned above, a sound standardization framework is a key factor for strengthening the national regulatory environment, at the same time providing an opportunity for interaction between government and academia. Such collaboration can lead, as in the case of Korea [22], to the actors working together in building a national "culture of standardization" that permeates all strata of society. The development of this culture requires a long-term vision with strategic, tactical and operational stages that include the participation of all stakeholders. Furthermore, reducing the standardization gap in a comprehensive manner calls for the adoption of an interdisciplinary approach.

2. Capacity building: An essential element for the deployment of a comprehensive national standardization plan, capacity building is particularly relevant for Health IT and related emerging disciplines. The importance of capacity building is evident when one considers that it is required for three of the four aspects of the Health IT ecosystem (see Fig. 3.1). Academic institutions play a key role in the design and implementation of educational and training programs that stress the technical, legal and innovation issues involved.
3. Learning by doing: The participation of Mexican specialists in ITU standardization working groups and in other international standards agencies would significantly contribute to the development of the skills and practical knowledge required in all the links of the Health IT value chain. Achieving this participation would, of course, require allocation of the necessary funds.

5 Final Remarks and Conclusions

Based on the analysis presented in Sections 2, 3 and 4, we identified major areas that should be addressed in order to improve the status of Health IT standardization in Mexico:

- The need to establish a national Health IT policy emphasizing the central role of standardization in the value chain of medical processes and technologies. The framework for Health IT standardization would be a key component of this national policy.
- Given the nature and dynamics of the Health IT ecosystem, the legal and ethical aspects of the framework are of particular importance and must be addressed.
- Capacity building and the establishment of a national program that permeates all levels of society are two critical aspects for fostering a culture of standardization in the country [22].
- As suggested by the WHO [16], the Health IT standardization framework should take into account the following disciplines: telemedicine, mHealth, eLearning in health sciences for all the medical specialties, management of patient and medical infrastructure, as well as the legal and ethical elements involved.
- In order to fully capitalize on interoperability, it is important to focus on patients' needs and their interaction with data and technology. In this respect, patient education and the departure from a fragmented and technology centric perspective are essential if we are to take advantage

of the full potential of Health IT. Our approach, which conceptualizes the Health IT ecosystem as a socio-technical-legal system, would support the development of an integrated and all-inclusive national Health IT plan.
- A perspective that is informed by complexity theory, as suggested in the flexible standards strategy [17], constitutes a pragmatic and integral approach for standards creation in developing countries. Complexity theory is proving to be relevant for understanding and explaining the challenges and needs of a sustainable Health IT framework.

An integral framework for Health IT standardization could provide a catalyst for moving away from a condition of mere technology usage toward a technology development model. Embracing this transition would allow, in the long term, the deployment of cost-effective medical services and products in order to meet the varying needs of citizens and health institutions.

Without a comprehensive framework for Health IT in developing countries, achieving the application of ICT as a vehicle for human development will remain elusive and an important opportunity to participate in the definition and creation of standards of global impact will be lost. The experiences of developing countries in the provision of Health IT services in rural and underserved areas are very important, and the development of a national standards framework for Health IT that is supported by all the regulatory agencies involved is, in our view, a fundamental step in the advancement of Mexico toward the knowledge society. Rather than defining or recommending the development of a particular standard or set of standards, our purpose in this article has been to propose a starting platform for Health IT standardization. We recognize that it is necessary to develop a strategic and tactical plan with its associated lines of action in which an approach from complexity science could contribute to understanding the context and interactions of all the actors involved in the Mexican health sector.

References

1. eHealth and Health Internet Domain Names. Report by the Secretariat, World Health Organization, 11 January 2013.
2. National eHealth Strategy Toolkit. World Health Organization and International Telecommunications Union. Available at: http://www.who.int/ehealth/en/
3. eHealth standardization and interoperability. WHO 132nd session, Agenda item 10.5, 28 January 2013.

4 Bridging the Standardization Gap: Measuring and Reducing the Standards Gap, ITU, 4 December 2009.
5 Tool for Assessing Standards Capability (TASC), ITU-T, 4 December 2009.
6 Promoting Access to Medical Technologies and Innovation, WHO-WIPO-WTO. Available at: http:// www.who.int/phi/en/
7 Krechmer, K. "Teaching Standards to Engineers", The International Journal of IT Standards and Standardization Research, Vol 5, No. 2, July-December 2007.
8 World Bank Data: Mexico. Available at: http://www.worldbank.org/country/mexico.
9 Manual del Expediente Clínico Electrónico, Dirección General de Información en Salud, Secretaría de Salud, México, 2011.
10 Serie Tecnologías en Salud, Volumen 3 Telemedicina, Secretaría de Salud, México, 2011.
11 Procesos Normativos en Salud, Dirección General de Calidad y Educación en Salud, Secretaría de Salud. Available at: http://www.calidad.salud.gob.mx/normatividad/procesos_normativos_salud.html
12 Centro Nacional de Excelencia Tecnológica en Salud, CENETEC. www.cenetec.gob.mx
13 Norma Oficial Mexicana NOM-024-SSA3-2010, Diario Oficial de la Federación, 16 Agosto 2010.
14 Ley Federal de Protección de Datos Personales en Posesión de los Particulares, Diario Oficial de la Federación, 5 Julio 2010.
15 El programa E-México, www.emexico.gob.mx
16 Global Observatory for eHealth, WHO, 2009. Available at: http://www.who.int/goe/data/en/
17 J. Braa et al, "Developing Health Information Systems in Developing Countries: The Flexible Standards Strategy", MIS Quarterly Vol. 31 Special Issue, pp. 1–22, August 2007.
18 Telemedicine. Opportunities and developments in Member States, Report on the second global survey on eHealth, Global Observatory for eHealth series, Volume 2, WHO, 2010.
19 N. Gertrudiz, "e-Health: the case of Mexico", Latin Am J Telehealth, Vol 2, No. 2, pp 127–167, 2010.

20 A. Velazquez, "The landscape of e-Health issues in Mexico" ETSI Meeting on eHealth, Sophie Antipolis 2007.
21 eHealth Strategy and Plan of Action (2012–2017), PAHO/WHO, 2011
22 Bum H. K. et al, Future Society & Standards, Korean Standards Association, 2004/2007.

Biographies

AMANDA O. GOMEZ-GONZALEZ Dr. Gomez-Gonzalez got her PhD in telecommunications. She has more than 20 years of experience regarding national and Latin American programs in telemedicine, telehealth and e-Health. She was founding director of the Mexican National Telehealth Programme. Dr. Gómez-González is currently governmental development manager of the Mexican Space Agency and Professor at the National Autonomous University of Mexico, UNAM.

VERONICA ROJAS-MENDIZABAL Veronica Rojas-Mendizabal is a Telecommunications Engineer. She earned her Master's degree in Telecommunications and Telematics from the Catholic University of Bolivia (UCB) in La Paz, Bolivia in 2008. She collaborated with the Regional Center for Education in Space Science and Technology for Latin America and the Caribbean (CRECTEALC), a United Nations space educational center, in Puebla, Mexico. She is currently a doctoral student in Electronics and Telecommunications at CICESE Research Center in Ensenada, Baja California, Mexico.

ROBERTO CONTE-GALVAN Dr. Roberto Conte-Galvan got his Ph.D. in Electrical Engineering from Virginia Polytechnic Institute & State University (Virginia Tech, USA) in 2000. He coordinated the first Gradute course on Telemedicine in Mexico and Latin America. He is currently a Full Researcher at the Department of Electronics and Telecommunications and member of the Telecommunications Networks Group at CICESE Research Center in Ensenada, Baja California, Mexico. Dr. Conte serves as representative and executive liaison with the Mexican Space Agency.

ANGELICA BAPTISTA SILVA Dr. Angelica Baptista Silva earned a PhD in Public Health in 2013. She is a specialist in Internet with a degree in Social Communication and Journalism. She was coordinator of the Health TV Channel at the Department of Information Technology of the Oswaldo Cruz Foundation/FioCruz from 1998-2011 and since 2012 coordinates the activities of the Telehealth Laboratory of the National Institute of Women, Children and Adolescent's Health Fernandes Figueira (IFF / FioCruz) and the International Network of Human Milk Banks. She is currently member of the Advisory Committee and coordinator of the Content Usage Working Group at the Brazilian Telemedicine University Network/RUTE.

ARTURO SERRANO-SANTOYO Dr. Serrano-Santoyo earned his Doctor's degree in Electrical Engineering from the National Polytechnic Institute in Mexico City in 1980. In 1981 he received the ALCATEL Annual Telecommunications Award for his contributions to rural satellite communications in Mexico and Latin America. He has been a telecommunications consultant for the Organization of American States and the United Nations, as well as for many private companies and governmental agencies. He is currently researcher at CICESE Research Center in Ensenada, Baja California, Mexico and Professor at the Autonomous University of Baja California.

Author Index

A. Sivabalan, 241, 251
Abbas Siddiqui, 329, 345
Amanda Gomez-Gonzalez, 347
Anand R. Prasad, 175, 184
Andreas Kunz, 109, 122
Angelica Baptista Silva, 347, 361
Arpan Pal, 253, 269
Arturo Serrano-Santoyo, 347, 361
Arvind Mathur, 205, 219
Ashok Chandra, 137, 157
Ashutosh Dutta, 187, 202
Henk J. de Vries, 25, 39
Hiroshi Nakanishi, 59, 79
JaeSeung Song, 109, 122
Katrine Bergh Andersen, 1, 23
Kohei Satoh, 271, 285
Kpatcha Bayarou, 329, 346
Krishna Sirohi, 221, 239
Kunio Igusa, 59, 80
Liljana Gavrilovska, 41, 57
M. A. Rajan, 241, 252
Mayur Dave, 175, 184
Munhwan Choi, 83, 107

Niels Madelung, 1, 23
Niranth Amogh, 187, 203
P. Balamuralidhar, 241, 252, 253
Parag Pruthi, 187, 201
Paul Müller, 329, 345
Pawan K. Garg, 137, 155
Purnendu S. M. Tripathi, 159, 173
Rahamatullah Khondoker, 329, 346
Ramjee Prasad, 123, 136
Ritesh Kumar Kalle, 187, 219
Robert M. vanWessel, 25, 39
Roberto Conte-Galvan, 347, 361
Rozhan Othman, 59, 80
Shozo Komaki, 59, 81
Soma Bandyopadhyay, 253, 267
Sunghyun Choi, 83, 107
Tilak R. Dua, 137, 156
Toshiaki Kurokawa, 287, 299
Veronica Rojas-Mendizabal, 347, 360
Vladimir Atanasovski, 41, 57
Weiping Sun, 83, 106

www.ingramcontent.com/pod-product-compliance
Ingram Content Group UK Ltd.
Pitfield, Milton Keynes, MK11 3LW, UK
UKHW021321180426
11947UKWH00015B/13/1